Trace Metal Adsorption Characteristics of Nanomaterials

Jared J. Gerschler

Universidad de la Cañada

Copyright © 2018 Jared J. Gerschler

All rights reserved.

ISBN: 1986241068
ISBN-13: 978-1986241069

CONTENTS

1.0 Introduction 5

1.1.0 Nanotechnology and nanomaterials 5

1.2.0 Carbon nanotubes 6
    1.2.1 Discovery and characterization 6
    1.2.2 Synthesis 6
    1.2.3 Properties and applications 7
    1.2.4 Toxicity 11
    1.2.5 Methods of surface oxidation 11
    1.2.6 Adsorption theory 12

1.3.0 Titanium dioxide 13
    1.3.1 Discovery and history 13
    1.3.2 Applications 15
    1.3.3 Synthesis 15
    1.3.4 Toxicity 25
    1.3.5 Adsorption theory 16
    1.3.6 Effects of particle size 16
    1.3.7 Effects of crystal structure 17

1.4.0 Antimony 17
    1.4.1 Environmental prevalence 17
    1.4.2 Applications 18
    1.4.3 Toxicity 18

1.5.0 Instrumentation for the analysis of nanomaterials 19
    1.5.1 Inductively coupled plasma – mass spectrometry (ICP-MS) 19
    1.5.2 Inductively coupled plasma – atomic adsorption spectrometry (ICP-AES) 21
    1.5.3 High performance liquid chromatography (HPLC) 21

    1.5.4 Ion exchange chromatography (IC) 22
    1.5.5 HPLC-ICP-MS coupling 22

1.6.0 Methods for the adsorption characterization of nanomaterials 23
    1.6.1 The Brunauer-Emmett-Teller (BET) method
        of surface area characterization 23
    1.6.2 Surface functionalization characterization
        using Boehm's titration 24
    1.6.3 The Freundlich and Langmuir isotherm models 25

1.7.0 Project Overview 28

2.0 Materials and Methods 30

2.1.0 Preparation of Sb(III) and Sb(V) solutions 31

2.2.0 Oxidation of multi-walled carbon nanotubes 31

2.3.0 Leaching of multi-walled carbon nanotubes 31

2.4.0 Assessment of the effect of adsorbent concentration
    on antimony adsorption at low pH 32

2.5.0 Investigating the effect of multi-walled carbon nanotube
    surface oxidation on Sb, Cd, and Pb adsorption capacity 33

2.6.0 Determination of the effect of substrate (Pb, Sb, and Cd)
    concentration on the adsorption potential of MWCNTs, AC,
    micro- and nanoparticle titanium dioxide 34

2.7.0 Investigating the antimony adsorption capacity of
    nanoparticle titanium dioxide 35

2.8.0 Investigating Sb(III) and Sb(V) adsorption by micro-
    and nanoparticle titanium dioxide; effects on speciation 36

2.9.0 Investigating the effect of light upon antimony (III) adsorption
    by nano- and microparticle titanium dioxide 36

2.10.0 Obtaining infrared spectra of n-TiO_2 and Sb
    adsorbed on n-TiO_2 37

2.11.0 Procedure for the analysis of solution samples
    by ICP-AES and ICP-MS 37

2.12.0 Procedure for the analysis of solution samples
by HPLC-ICP-MS... 38

2.13.0 Fitting adsorption data to linear, Langmuir, and
Freundlich isotherms 39

3.0 Results and Discussion... 40

3.1.0 Leaching of multi-walled carbon nanotubes 40

3.2.0 Assessment of the effect of adsorbent concentration
on antimony adsorption at low pH 42

3.3.0 Investigating the effect of multi-walled carbon nanotube
surface oxidation on Sb, Cd, and Pb adsorption capacity 43

3.4.0 Determination of the effect of substrate (Pb, Sb, and Cd)
concentration on the adsorption potential of MWCNTs, AC,
micro- and nanoparticle titanium dioxide 51

3.5.0 Investigating the antimony adsorption capacity
of nanoparticle titanium dioxide 68

3.6.0 Investigating Sb(III) and Sb(V) adsorption by micro-
and nanoparticle titanium dioxide; effect of light, effects on
speciation 69

3.7.0 Infrared spectra of n-TiO_2 and Sb adsorbed on n-TiO_2 85

3.8.0 Summary 87

4.0 Conclusion 88

4.1.0 Future Work 90

5.0 References 91

6.0 List of Tables 100

7.0 List of Figures 101

JARED J. GERSCHLER

ACKNOWLEDGMENTS

niná xi tajbechili nga kabisekaonain Nae

1.0 INTRODUCTION

1.1.0 Nanotechnology and Nanomaterials

Nanotechnology is the study of materials on the atomic and molecular scale. The physical and chemical behavior of a particle is often a function of its size, and it has been found that materials on the nanoscale behave in a manner often distinctly different from the behavior of the bulk material (Brumfiel, 2003). By manipulating materials at the nanoscale, potentially useful properties may be exhibited. Nanotechnology is a multidisciplinary field, and is applicable across a wide range of disciplines including chemistry, biology, and physics.

Nanotechnology has already led to the creation of a wide range of new materials and devices. Breakthroughs in energy storage, filtration, and cancer treatments are but a few of the novel uses of nanomaterials. Already, the market for nanomaterials and nanotechnology is growing. For the foreseeable future, the demand for products based on nanotechnology is expected to drastically increase (NSF, 2001).

Of course, with all of the promise shown by nanotechnology, a certain amount of skepticism is also necessary. The rate of nanotechnological development is rapid, and the global impact of nanotechnology is still not completely understood. Already, studies have shown that some nanomaterials, such as carbon nanotubes (CNTs), exhibit cellular toxicity and also impact lung function (Stern and McNeil, 2008).

The rapid growth of nanotechnology, coupled with the potential toxicity of some nanomaterials, emphasizes the importance of more research concerning nanomaterial bioavailability and environmental fate. Furthermore, the toxicity of a nanoparticle is also a function of its adsorbed components. In this text, we investigate the adsorption of antimony and other trace elements by multi-walled carbon nanotubes (MWCNTs), activated carbon (for comparison to MWCNT adsorption), micro- and nanoparticle titanium dioxide.

1.2.0 Carbon nanotubes

1.2.1 Discovery of carbon nanotubes

Carbon nanotubes were first reported in the literature in 1952 by Radushkevich and Lukyanovich in the Russian Journal of Physical Chemistry (Monthioux and Kuznetsov, 2006). The discovery originally went unnoticed by western scientists, as the journal was published chiefly in Russian. In 1976, however, Oberlin et al. reported the growth of carbon filaments as a result of benzene decomposition (Oberlin et al., 1976). This paper also failed to cause any large amount of excitement in the scientific community, and in fact it was not until 1991, with Iijima's report of "Helical microtubules of graphitic carbon" in the journal Nature, that carbon nanotubes began to be seriously explored (Iijima, 1991).

1.2.2 Synthesis of carbon nanotubes

There are currently a number of methods used to synthesize carbon nanotubes. While all have potential advantages and disadvantages, the search is still underway for more efficient methods of CNT production.

Arc discharge was the first major method used to synthesize carbon nanotubes. The process involves creating an arc discharge in an inert atmosphere by passing a large current (~100 Amps) through a pair of graphite electrodes. Carbon soot produced by the cathode during the arc discharge process may contain up to 30 percent CNTs by weight. Both single- and multi-walled carbon nanotubes with length up to 50 μm are created using this technique (Collins and Phaedon, 2000). A newer technique, laser ablation, uses a series of laser pulses to vaporize graphite in a high temperature reactor. An inert gas such as argon is bled into the chamber, and the nanotubes develop by condensing on the cooler surfaces of the reactor. Laser ablation tends to produce single-walled carbon nanotubes (SWCNTs); the diameter of the tubes is dependent upon the reactor temperature. Total yield is approximately 70% (Collins and Phaedon, 2000).

Chemical vapor deposition (CVD) is another method used to synthesize carbon nanotubes. A layer of catalytic metal particles (Ni, Co, and Fe are the most common) are applied to a substrate surface such as silicon. The size of the nanoparticles will affect the final diameter of the CNTs (Ebbesen et al., 1992). The substrate is then heated to a high temperature (~700 °C), and two gasses are released into the development chamber. One gas is a process gas (usually ammonia, nitrogen or hydrogen), and the other must be a carbon containing gas, such as acetylene or ethylene. As the gases enter the chamber, carbon from the carbon containing gas is deposited on the catalytic particles. Chemical vapor deposition can be used to synthesize both MWCNTS and

SWCNTs (Kong, et al. 1998). The nanotube growth mechanism is known to depend upon temperature and catalyst-substrate interactions, and in the case of microwave plasma CVD, CNTs are found to grow via a "base growth" mechanism. The kinetics of the "base growth" mechanism involve a very rapid initial growth, followed by a period of stunted growth–the so-called "saturation period." CNTs synthesized via CVD tend to incorporate a section of the original growth catalyst at one end of the nanotube, and there is a correlation between the catalyst layer thickness and the diameter of the nanotubes (Li et al., 1996).

While carbon nanotubes are primarily manmade in origin, research has indicated that they also form in small amounts during the combustion of a paraffin wax candle (Li and Hseih, 2007). Also, common soot often contains small concentrations of carbon nanotubes; they are also generated during the combustion of carbon-rich substances such as methane, ethylene, and benzene (Duan and McKinnon, 1994; Yuan et al., 2001). Extraterrestrial carbon has also been shown to contain carbon nanotubes and fullerene-like structures (Basiuka et al., 2005).

1.2.3 Properties and applications of carbon nanotubes

Many of the properties of carbon nanotubes are influenced by their structure; as previously mentioned, there exist both single-walled carbon nanotubes (SWCNTs) and multi-walled-carbon nanotubes (MWCNTs). There are a wide variety of carbon nanotube structures, including, in the case of SWCNTs, the "armchair" and "zig-zag" structures (see Figs. 1.2a – 1.2c). Armchair nanotubes are metallic in nature, while zig-zag nanotubes possess semiconducting properties. The difference in properties in SWCNTs results from a difference in structure, which can be imagined as resulting from a different angle with which a graphene sheet can be rolled up to form the nanotube. The tensile strength of armchair SWCNTs is expected to be greater than that of zig-zag SWCNTs. The tensile strength of MWCNTs is theorized to be greater than that of any SWCNTs (Dresselhaus et al., 2000).

There is an extremely large variety of uses for carbon nanotubes. Some uses have already been realized, while others are still in development. The majority of these applications rely on either the high tensile strength or the unique electronic and chemical properties of CNTs.

Carbon nanotubes are characterized by their immense tensile strength, and are unique in combining both strength and flexibility (Popov, 2004). Along the length axis, carbon nanotubes possess a greater tensile strength than steel. This high tensile strength to weight ratio suggests clothing applications, and indeed, the Institute for Soldier Nanotechnologies at the Massachusetts Institute of Technology (MIT) is already working on the development of an extremely rugged combat jacket based on CNTs (MIT

ISN, 2008). The use of carbon nanotubes in a bulk material has been found to increase the tensile strength of the bulk material. When carbon nanotubes are added to concrete, the increase in tensile strength helps to halt the propagation of cracks. The formation of a composite created by adding CNTs to polyethylene has been found to increase the elastic modulus of the polyethylene by approximately 30% (Tang et al., 2003). On a larger scale, the use of CNTs in the construction of suspension bridges has been suggested. A more exotic application imagined is that of a space elevator; a CNT-based tether could provide a facile method of moving objects into low earth orbit.

Carbon nanotubes also possess unique electrical properties, partially a result of the large network of sp2 hybridized carbon atoms, allowing for a wide range of uses. CNTs have been used to create artificial muscles (Baughman et al., 1999), conductive films for use in displays, and as components in computer circuits (Collins and Phaedon, 2000). In the arena of electronic engineering, CNTs have been used to create nanowires, which may one day lead to the shrinking of integrated circuits (Zhang et al., 2000). Furthermore, ultracapacitors have been constructed by binding CNTs to the charge plates of normal capacitors to increase the surface area (Tang et al., 2001). Additionally, field effect transistors (FETs) have been fabricated using CNTs, and these CNT-FETs possess superior operational speed and lower power consumption when compared to traditional metal oxide semiconductor FETs (MOSFETs) (Guo et al., 2002). Lastly, there are myriad medical applications for carbon nanotubes. Since CNTs have been shown to be easily transported into cells, some biochemical research has centered on using CNTs as drug delivery systems. Additionally, anticancer research has even investigated the use of radio frequency radiation to destroy CNT-containing cancer cells.

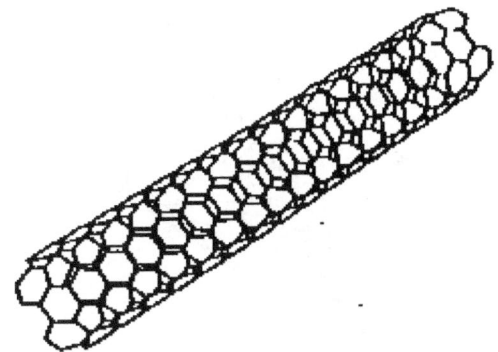

Figure 1.2a. Armchair nanotube (10,10). Metallic. Image courtesy of the World Technology Development Center of Loyola College. (http://wtec.org)

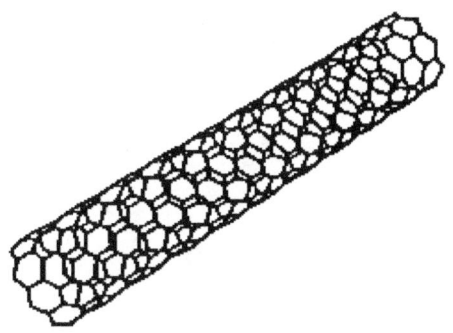

Figure 1.2b. Zig-zag nanotube (0,10). Semiconducting. Image courtesy of the World Technology Development Center of Loyola College. (http://wtec.org)

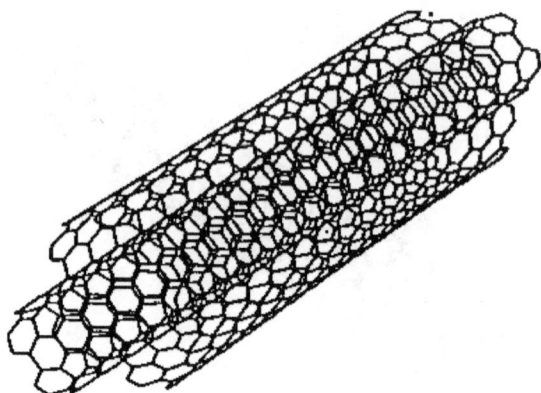

Figure 1.2c. Multi-walled carbon nanotube. Image courtesy of the World Technology Development Center of Loyola College. (http://wtec.org)

Carbon nanotubes have been shown to possess good adsorption properties for a range of substances, both gases and solution-phase elements and compounds (Chang et al., 2001; Chen et al., 2007). Surface-oxidized CNTs have been shown to have an affinity for the public water contaminants arsenic and lead, and their incorporation into filtration systems is currently being implemented (Dreher, 2004; Chowdury et al., 2000; Peng et al., 2003). Before such filtration systems are widely utilized, however, the interactions of CNTs with the environment must be studied in depth.

1.2.4 Carbon nanotube toxicity

The toxicity of carbon nanotubes has been the subject of several studies (Murr, 2007). It is widely accepted that MWCNTs easily pass through cell walls, and are transported into cytoplasmic vacuoles with facility (Monteiro-Riviere et al., 2005). Upon investigating the effect of multi-walled carbon nanotubes on T lymphocytes, Bottinni et al., (2006) found that MWCNTs were capable of inducing cellular apoptosis, but only at rather high MWCNT concentrations (~400 μg/mL).

The effect of carbon nanotubes on macroorganisms has primarily been investigated using rats. Inhalation of CNTs by rats tends to result in the production of transient inflammatory and cell injury effects. Warheit et al. (2004) reported the formation of epithelioid granulomas, and came to the conclusion that carbon nanotubes are much more toxic than bulk carbon black, and may also be more toxic than nanoparticle quartz.

In general, the toxicity of carbon nanotubes is not exclusively a result of CNT size, and may be augmented by toxins adsorbed to the nanoparticle (Colvin, 2003). Because of these toxicity risks, combined with the growing market for nanomaterials, it is of utmost importance that the impacts of nanomaterials on the environment be studied (Linkov et al., 2007, and Helland et al., 2008).

1.2.5 Methods for surface oxidation of carbon nanotubes

Surface-oxidized CNTs often show superior adsorption characteristics when compared to pristine CNTs. There are currently several widely used methods for performing surface oxidation of carbon nanotubes. The use of potassium permanganate to perform surface functionalization of CNTs is well-established. In this method, the nanotubes are heated in a solution containing potassium permanganate at either an alkaline or acidic pH. An acidic solution of hydrogen peroxide, when heated, has also been used to perform the surface oxidation of CNTs. Finally, using heated (80-140 °C), concentrated nitric acid has been utilized as an oxidation method (Aitchison et al., 2007), although reaction time may greatly vary depending on experimental conditions. Inherent in all surface oxidation methods is some degree of CNT decomposition (Li et al., 2003). Typically, the stronger the oxidizer and the longer the reaction period, the greater the decomposition.

1.2.6 The theoretical basis for adsorption by carbon nanotubes

The mechanism of adsorption by pristine carbon nanotubes is primarily based on physisorption (Kocabas et al., 2008). Physisorption (physical adsorption) is the result of van der Waals interactions between the adsorbent and adsorbate (Wang et al., 2007). Such interactions are weak in nature (compared to ionic interactions, for example), occur quickly and are reversible; they possess no activation barrier (Chen et al., 2008). They are surface symmetry insensitive, and allow for the multilayer formation of adsorbates (Piao et al., 2008).

Physisorption is the well-established mechanism of adsorption onto CNTs when the adsorbate is gaseous (such as NH_3 or NO_2) or a solute (Cataldo et al., 2008; Chang et al., 2001). The mechanism of adsorption of carbon nanotubes that are not pristine (those that have been modified through a surface functionalization process) is not simple physisorption (Chen et al., 2007). Cationic metal species, such as Pb(II), are adsorbed to the surface of surface-oxidized MWCNTs through a mechanism involving complex formation with the oxygen-containing functional groups that are introduced during the oxidation process (Stafiej et al., 2007; Li et al., 2002).

1.3.0 Titanium dioxide

1.3.1 The discovery and history of titanium dioxide

The discovery of titanium dioxide took place in 1791, when William Gregor isolated a white metallic oxide from magnetite-containing sands in England. Titanium dioxide was not used commercially until 1916, when it was initially used as a white pigment.

Titanium dioxide is a naturally-occurring mineral, primarily found in its most stable form, rutile. There are other forms of titanium dioxide known (the monoclinic baddeleyite and the orthorhombic α-PbO_2 analogs), but only three occur in significant amounts: rutile, anatase, and brookite. Both anatase and brookite will convert to rutile with heating (Greenwood and Earnshaw, 1997). Additionally, rutile is also the densest of the structural analogs of titanium dioxide and strongly absorbs radiation in the ultraviolet region of the electromagnetic spectrum. As such, it is used as the active ingredient in some sunscreens. In terms of photocatalytic ability, nanocrystalline anatase exhibits the highest activity. Brookite is not photoactive. Both anatase and rutile possess tetragonal crystal structures, while that of brookite is orthorhombic (see Figs. 1.3a – 1.3c).

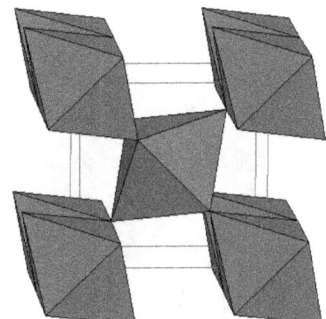

Figure 1.3a. Crystal structure of rutile. Tetragonal. Image courtesy of the University of Colorado at Boulder Mineralogy Program.

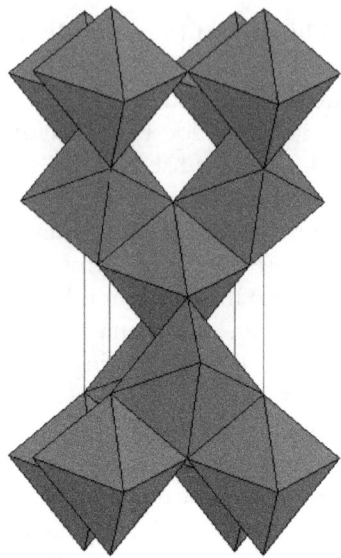

Figure 1.3b. Crystal structure of anatase. Tetragonal. Image courtesy of the University of Colorado at Boulder Mineralogy Program.

Figure 1.3c. Crystal structure of brookite. Orthorhombic. Image courtesy of the University of Colorado at Boulder Mineralogy Program.

1.3.2 Applications of nanoparticle titanium dioxide

Applications of nanoparticle titanium dioxide typically revolve around either its oxidative or semiconducting properties. Nanoparticle titanium dioxide has been used in paints as both a whitening agent and an antifungal. In the presence of ultraviolet light, it is capable of hydrolyzing water into hydrogen and oxygen. Dye-sensitized photovoltaic cells utilize titanium dioxide for its semiconducting properties; it has been found that nanoparticle titanium dioxide is much more effective in photovoltaic and photoelectrochemical cells when compared to non-nanostructured titanium dioxide. Additionally, it is used for a wide variety of chemical sensors (Varghese et al., 2003 and Reddya et al., 2002).

1.3.3 The synthesis of nanoparticle titanium dioxide

Although titanium dioxide is naturally-occurring, the production of nanocrystalline titanium dioxide is typically performed in a laboratory setting. Synthesis procedures usually involve the hydrolysis of a Ti(IV) salt, such as titanium tetrachloride or titanium tetrabutoxide (Kolenko et al., 2003). By varying the conditions (temperature, concentration) of the hydrolysis, the particle size may be affected. Such titanium (IV) salts may be fed into a diffusion flame reactor to gain greater control over particle size (Jang and Kim, 2001). Nanocrystalline anatase powders with particle sizes from 8 to 38 nm have been synthesized by Kalenko et al (2003) by the hydrothermal treatment of aqueous titanyl sulfate solutions. More recently, Gao et al. (2004) have reported a novel nanocrystalline titanium dioxide synthesis using an aqueous peroxotitanate solution prepared from H_2TiO_3, ammonia and H_2O_2.

1.3.4 The toxicity of nanoparticle titanium dioxide

It has been found that a high concentration (~100 μg/mL) of nanoparticle titanium dioxide is necessary before a cell begins to exhibit symptoms of toxicity. Interestingly, it appears that the toxicity of a titanium dioxide nanoparticle is not dependent on its surface area or comparative size. Interestingly, however, the phase composition of nanoparticle titanium dioxide is of supreme importance in determining toxicity. Anatase titanium dioxide was found to be approximately 100 times more toxic that rutile titanium dioxide. Generally, the most cytotoxic nanoparticles are also the most effective at generating reactive oxygen species (Sayes et al., 2006).

Rainbow trout (Oncorhynchus mykiss) exposed to nanoparticle titanium dioxide concentrations between 100 ng/mL and 1 μg/mL exhibited edema of the gills and thickening of the lamellae. Additionally, changes in the concentration of zinc and copper, particularly in the brain, were evident.

Overall, researchers have determined that titanium dioxide nanoparticles are not a major ionoregulatory toxicant (Federici et al., 2007).

Rats exposed to aerosolized nanoparticle titanium dioxide at low (10 – 50 mg/m^3) concentrations show no adverse health effects. When the atmospheric concentration of nanoparticle titanium dioxide is raised to 250 mg/m^3, however, extended exposure leads to the formation of bronchioloalveolar adenomas and cystic keratinizing squamous cell carcinomas. The lung tumors are different from human lung cancers, in terms of anatomic location, and do not exhibit tumor metastasis (Lee et al., 1985).

1.3.5 Theoretical description of nano- and microcrystalline titanium dioxide adsorption capacity

General studies have found that the adsorption capability of titanium dioxide is a result of the large number of hydroxyl groups present on the surface of titanium dioxide particles (Zhang et al., 2008). Although a specific study investigating the adsorption mechanism of antimony species and titanium dioxide has not yet been performed, an exact analogue has been completed using arsenic (Pena et al., 2006). Arsenic, being located directly above antimony in the periodic table, shares many chemical properties. Similar to arsenic, antimony is also toxic to most biological organisms.

Studies involving the adsorption of arsenic (III) and (V) onto nanocrystalline titanium dioxide have shown that both arsenic species form negatively charged inner-sphere complexes with the surface of the titanium dioxide (Pena et al., 2006). The complexes were investigated using extended x-ray absorption fine structure (EXAFS) spectroscopy. It was determined that the surface complexes on TiO$_2$ maintained the same non-protonated speciation at pH values from 5 to 10, and the dominant surface species were $(TiO)_2AsO_2$ and $(TiO)_2AsO^-$ for As(V) and As(III), respectively. Nanoparticle titanium dioxide was found to be a good sorbent of arsenic, due to the high surface area of the particles, and the large number of high-affinity surface hydroxyl groups (Pena et al., 2006). Provided the strong adsorption of arsenic by titanium dioxide, it should be expected that antimony species would also be easily adsorbed. Furthermore, it is suspected that the antimony species present at the titanium dioxide surface would be the analogues of the arsenic species.

1.3.6 The effect of titanium dioxide particle size on adsorption

The adsorption capability of titanium dioxide particles is largely an effect of particle size. In general, for most adsorbents, smaller particles possess a higher total surface area, and thus binding area. In the case of arsenic adsorption by magnetite particles, for example, higher adsorption capacity is known to correlate with decreased particle size (Yean et al., 2005). This holds true in the case of titanium dioxide; Jang et al., (2001) studied the effect of particle size on the photocatalytic properties of TiO_2. The rate of decomposition of methylene blue was found to be inversely related to particle size. This is to be expected, since the adsorbate (methylene blue) must first undergo sorption to the surface of the titanium dioxide particle before photodecomposition can occur. The smaller the titanium dioxide particle, the larger the catalytic surface area, and thus the higher rate of substrate decomposition. Giammar et al. (2007) explored the adsorption of lead to titanium dioxide of various particle sizes and crystallinities. The titanium dioxide nanoparticles studied had a diameter ranging from 20 - 33 nm, while the bulk titanium dioxide particles possessed a mean diameter of 520 nm. It was found that on a mass basis, adsorption by the titanium dioxide nanoparticles was always higher than that of bulk titanium dioxide. When the adsorption was normalized to surface area, however, the bulk material had a higher adsorption capacity.

1.3.7 The effect of crystal structure on titanium dioxide adsorption capacity

Pure anatase has been found to be, in general, a better adsorbent than rutile. Anatase containing even a small amount of rutile will perform inferiorly when compared to the pure substance. A comprehensive comparison between anatase, rutile, and brookite adsorption capacities has not been made.

1.4.0 Antimony

1.4.1 Environmental prevalence of antimony

Antimony is not considered an abundant element, but it is found in over 100 minerals. It most commonly exists in the -3, +3 and +5 oxidation states. The most common mineral form of antimony is stibnite (Sb_2S_3), which also happens to be the most common mineral used for ore. Other minerals include Valentinite (Sb_2O_3), Roméite (($Ca,Fe^{2+},Mn,Na)_2(Sb,Ti)_2O_6(O,OH,F)$) and Breithauptite (NiSb). Antimony-producing countries include China, South Africa, Bolivia, and Tajikistan (Chen and Yang, 2007).

1.4.2 Applications of antimony

Antimony is used widely in flame retardants and automotive brake pads, and has also found increasing use in the semi-conductor industry, being utilized in diodes and infrared detectors (Van Velzen and Langenkamp, 1998). Recently, antimony-doped tin oxide films have been used in conjunction with carbon nanotubes to create transparent conducting films (Castro and Schmidt, 2007). Such films are not only desirable for being both transparent and conductive, but also have the potential to be used as chemical sensors (Liu et al. 2008). Antimony is also used as a hardener in lead for batteries and bullets—it is found in the soil of shooting ranges, where Sb(V) is the dominant species (Steely et al., 2007). It is used in some cable sheathing, and also in lead-free solder. In the past, antimony containing drugs were used to treat Schistosomiasis. Polyethylene terephthalate (PETE) plastic is synthesized with the help of an antimony catalyst.

1.4.3 Antimony toxicity

The toxicity of antimony compounds is well-documented (Winship, 1987; Newton et al., 1976). In general, trivalent antimony compounds are more toxic than pentavalent compounds; this is partially a result of the increased aqueous solubility of Sb(III) species (Gebel et al., 1997). Oral LD50 values for studies on rats range from 115 mg/kg (antimony potassium tartrate) to approximately 20,000 mg/kg (antimony (III) oxide). Effects of exposure to any antimony (III) compound have been shown to result in nausea, vomiting, weakness, abdominal colic, diarrhea, and skin rashes (Chen and Yang., 2007). Symptoms of antimony exposure may also be exhibited by the cardiopulmonary system. Cough and chronic lung changes are evident in animals exposed to antimony (Oorts et al., 2008). Cardiotoxicity may result in arrhythmias, myocardial depression and damage, heart failure, and cardiac arrest (Newton et al., 2004). Antimony compounds have been shown to be potentially toxic to the mammalian reproductive system, and have mutagenic, and oncogenic potential. Chromosome damage has been reported (Girgis et al., 1965).

Interestingly, simple organisms have been shown to be capable of developing resistance to antimony species. Escherichia coli and Staphylococcus aureus are two bacteria which have been shown to develop a plasmid-induced resistance to both trivalent and pentavalent antimony (Silver et al., 1981).

1.5.0 Instrumentation for the analysis of nanomaterials associated with trace metals

1.5.1 Inductively coupled plasma – mass spectrometry (ICP-MS)

Inductively coupled plasma – mass spectrometry is a technique used for the analysis of a range of metals (and some non-metals) at low concentrations. Samples in solution or slurry form are introduced into an ionizing plasma. Following atomization and ionization of the sample, it is directed into a mass spectrometer, which sorts ions based on their mass-to-charge (m/z) ratios. The ion intensity at a given mass-to-charge ratio is then used to derive the concentration of the analyte.

A variety of methods are used to introduce samples into the ICP-MS. Electrothermal vaporization (ETV), laser ablation (LA), and aerosolization using a liquid nebulizer (LN) are all typical approaches for sample introduction; the use of a nebulizer is the most common method for liquid introduction. In this method, the sample is supplied as a solution or very fine slurry, and converted from a liquid into an aerosol within the nebulizer. The aerosol and carrier gas are then fed into the central channel of the argon plasma.

The inductively coupled plasma used in ICP-MS is created by the ionization of argon gas. The plasma is sustained in a torch, usually made of quartz, whose distal end is placed inside of an induction coil. A radio-frequency (RF) electric current is supplied to the induction coil. A spark is used to provide a source of electrons, which are then accelerated within the magnetic field of the induction coil. These electrons initiate the argon ionization

$$Ar^0 \rightarrow Ar^+ + e^-$$

The reverse reaction is also possible, and in practice an equilibrium is reached in which the rate of argon ionization is equal to the rate of recombination of electrons with argon ions. The result is an intense, hot (electron temperatures of 7,000 – 10,000 K) plasma which can easily atomize and ionize a wide variety of compounds to their substituent elements.

After exposure to the plasma, ionized analyte atoms are extracted through sampler and skimmer cones (which help to maintain the vacuum that is required by the mass spectrometer), and into the mass spectrometer for analysis. Ions (M^+) are sorted according to mass/charge ratio by a quadrupole mass analyzer. The instrument is calibrated for test elements using a series of known multielemental standards at the appropriate concentration range.

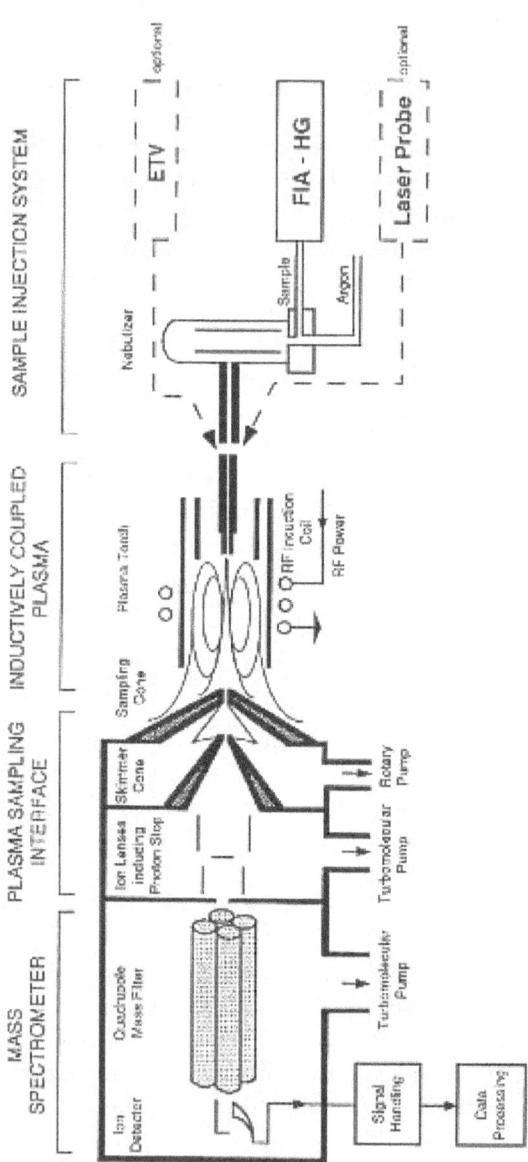

Figure 1.5a: Diagram of an inductively coupled plasma – mass spectrometer. Image courtesy of the University of Rhode Island, 2008.

1.5.2 Inductively coupled plasma – atomic emission spectroscopy (ICP-AES)

Inductively coupled plasma-atomic emission spectroscopy (ICP-AES) is similar to ICP- MS, in that an argon plasma is used to ionize samples before analysis. Furthermore, a nebulizer is usually used to introduce the samples to the plasma. The plasma vaporizes, atomizes, ionizes the sample, and excites electrons from their ground state. As the excited atoms return to the ground state, electromagnetic radiation is emitted at a wavelength characteristic of each element; this radiation provides a spectroscopic signature which may be used to identify the element. A shear gas is used to cut the plasma flame axially or radially, and at least one transfer lens is used to direct the emitted light upon a diffraction grating. The diffraction grating separates the emitted light into its individual components, which are then analyzed by photomultiplier tubes, a photodiode array, or a charge-coupled device (CCD). The intensity of light at a given wavelength is then compared to the intensity that was found for known concentrations of a suite of known elemental standards, and the concentration in the unknown may then be determined.

1.5.3 High-performance liquid chromatography (HPLC)

High-performance liquid chromatography (HPLC) is typically used for the separation, identification and quantification of chemical compounds. There are a wide range of modified HPLC techniques: normal phase chromatography, reversed phase chromatography, size exclusion chromatography, ion exchange chromatography and bioaffinity chromatography.

A typical HPLC setup consists of a mobile phase, a stationary phase, a detector, and a computer system to display the results. The stationary phase is typically a column packed with a polar, ionic or other material depending on the compounds that will analyzed. A pump is used to move the mobile phase (and the compounds contained therein) through the stationary phase (column). As compounds elute from the column, they are distinguished by a detector (often a UV absorption detector). The intensity of the absorption is partially determined by the concentration of the compound in the mobile phase. The retention time of the compound is a result of the interactions between the stationary phase and the mobile phase, the interactions between the compounds being analyzed and the column, and the composition of the solvent being used.

1.5.4 Ion exchange chromatography (IC)

Ion exchange chromatography is an HPLC method in which a column containing ionic functional groups is used as the stationary phase. The charged functional groups on the column interact with analytes of the opposite charge. Both anionic and cationic ion exchange columns exist; antimony (III) and (V) studies utilize an anion exchange column.

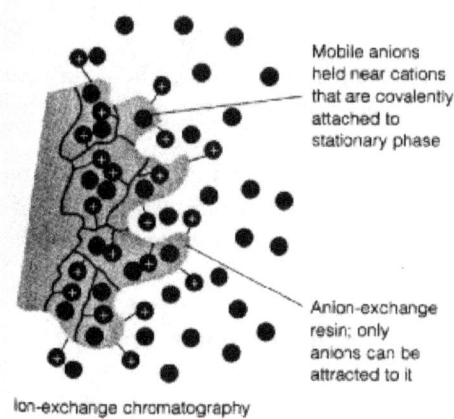

Figure 1.5b: Ion exchange chromatography. Image credit: Rensselaer Polytechnic Institute, Chemical Engineering Department, 2008.

During traditional operation, a buffered mobile phase is used to introduce the sample into the column. The analyte being studied will be retained on the stationary phase, and eventually be removed by the mobile phase and transported to the detector. Retention time (t_r) is a result of the strength of analyte – stationary phase ionic interactions. The stronger the interaction, the longer the retention time.

1.5.5 ICP-MS and HPLC coupling

For speciation work, HPLC and ICP-MS may be coupled together. The HPLC is fitted with an ion exchange column, and the outflow of the chromatograph is connected to the nebulizer on the ICP-MS (Newton et al., 2006 and Ruiz-Haas et al., 1998). The analyte is then introduced to the HPLC, and as analyte elutes off of the column, it is directed to the ICP-MS. If a calibration curve has been created using different concentrations and analytes of species, the data provided by the ICP-MS may be used to provide a quantitative analysis of the analyte present, as well as its speciation.

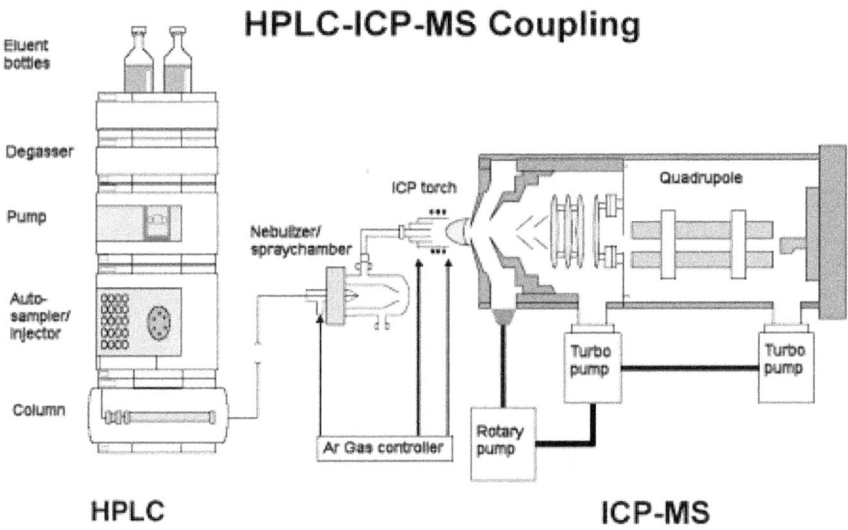

Figure 1.5c: HPLC coupled with ICP-MS. Image Credit: European Virtual Institute for Speciation Analysis, 2008

1.6.0 Methods for the adsorption characterization of nanomaterials

1.6.1 The Brunauer-Emmett-Teller (BET) method of surface area characterization

Brunauer-Emmett-Teller (BET) theory is a model explaining the adsorption of gas molecules on solid surfaces (Brunauer et al., 1938). The basis of the theory is an extension of Langmuir theory; it is assumed that gas molecules may form an infinite number of adsorption layers upon the original substrate, and that each molecular layer may be described by Langmuir theory. There is assumed to be no interaction between separate layers. The traditional form of the BET equation is

$$\frac{1}{v[(\frac{P_0}{P})-1]} = \frac{c-1}{v_m c}\left(\frac{P_0}{P}\right) + \frac{1}{v_m c} \qquad (1)$$

where P and P_0 are the equilibrium and saturation pressure of the adsorbates (varying with temperature), respectively. The BET constant is represented by c; the total adsorbed amount of gas is indicated by v, and v_m is the amount of gas adsorbed on a single monolayer. BET adsorption experiments are an important method used to determine the specific surface area of a material. For linear approximations of the BET equation, the total and specific surface area ($S_{BET, Total}$ and S_{BET}, respectively) of a solid may be determined using the following equations:

$$S_{BET,Total} = \frac{v_m N_a s}{V} \qquad (2)$$

$$S_{BET} = \frac{S_{Total}}{b} \qquad (3)$$

Where N_a is Avogadro's number, s is the adsorption cross section (i.e. probability of successful adsorption of the adsorbate), V is the molar volume of the adsorbent gas, v_m is the quantity of gas adsorbed by a single monolayer, and b is the molecular mass of the adsorbed species (Jiaa et al., 2007).

1.6.2 Surface functionalization characterization using Boehm's titration

Boehm's titration is a popular method for determining the functionalization introduced on CNT's during the oxidation process (Boehm et al., 1964). In a typical procedure, the CNTs are immersed into a series of solutions, traditionally 0.05 M HCl, $NaHCO_3$, Na_2CO_3, and NaOH. The number of basic sites present on the CNTs is calculated by determining the amount of HCl that has reacted. Additionally, the number of free acidic groups can be determined by titration with $NaHCO_3$, Na_2CO_3, and NaOH. It is assumed that NaOH neutralizes carboxyl, lactone (neutralized via a hydroxide-induced ring opening) and phenolic groups, Na_2CO_3 neutralizes carboxyl and lactone groups, and $NaHCO_3$ neutralizes only carboxylic groups.

1.6.3 The theoretical basis for the Langmuir and Freundlich adsorption models

The Langmuir and Freundlich adsorption models were originally developed to describe the adsorption and desorption of gases on solid surfaces.

The Langmuir model for adsorption is a simplistic model composed of several key assumptions (Yang, 1998). First, it is assumed that adsorption cannot exceed monolayer coverage; i.e. that an adsorbate molecule cannot adsorb to other adsorbate molecules. The next assumption is that the surface of the adsorbent is uniform, and that all adsorption sites are equal. Finally, it is required that the ability of a molecule to adsorb to a given site is not reliant on the occupation of neighboring sites.

$$q = \frac{q_m K_L C}{1 + K_L C} \quad (4)$$

In equation (4), q is the concentration of the adsorbed analyte (mg/g) and C is the equilibrium analyte concentration (the concentration of analyte in solution once the adsorbent/adsorbate system has reached equilibrium) (mg/L). The Langmuir constants q_m and K_L are related to the adsorbent adsorption capacity and energy of adsorption, respectively. A linear form of the Langmuir equation can be derived from (4), and allows a fit to the data to be easily found:

$$\frac{1}{q} = \frac{1 + K_L C}{q_m K_L C}$$

$$\frac{1}{q} = \frac{1}{q_m K_L C} + \frac{K_L C}{q_m K_L C} \quad (5)$$

$$\frac{1}{q} = \frac{1}{q_m K_L} \frac{1}{C} + \frac{1}{q_m}$$

A plot of 1/q versus 1/C then allows for extraction of the Langmuir parameters.

The Freundlich isotherm is a slightly more complex isotherm (Freundlich, 1926). Unlike the Langmuir equation, the empirical form of the

Freundlich equation is not limited to the description of solely monolayer adsorption. It does not approach a fixed adsorption capacity, however, as the concentration of analyte reaches saturation capacity of the adsorbent (Chiou, 2002).

$$q = K_F C^{\frac{1}{n}} \tag{6}$$

As in the case of the Langmuir equation (6), q is the concentration of the adsorbed analyte (mg/g), and C is the equilibrium solution analyte concentration (mg/L). K_F and n are called the Freundlich constants, and are related to the adsorption capacity and heat of adsorption, respectively. The Freundlich equation, in contrast to the Langmuir equation, does not approach arithmetic linearity at low concentrations, and can be made linear through the use of logarithms:

$$\ln(q) = \ln K_F + \ln C^{\frac{1}{n}}$$

$$\ln(q) = \frac{1}{n} \ln C + \ln K_F \tag{7}$$

A log plot then allows for the extraction of n and K_F.

The linearization of the Langmuir and Freundlich equations is typically used to allow equation parameters to be extracted using simple linear regression. While using the linear equation forms greatly simplifies the fitting process, it has been found to limit the accuracy of the determined Langmuir and Freundlich parameters (Bolstera and Hornbergerb, 2007). Because the Langmuir and Freundlich isotherms were originally modeled after systems in which the adsorbate was gaseous, the models work best to describe physisorption (as opposed to chemisorption) processes.

Table 1.6: Physicochemical significance of Langmuir and Freundlich parameters*

Langmuir Equation Parameter	Significance
q_m	Monolayer adsorption capacity of adsorbate (i.e. when surface of the adsorbent is completely covered in a monolayer of adsorbate).
K_L	Related to the heat of adsorption per unit mass of adsorbate; this is independent of the adsorbed amount.

Freundlich Equation Parameter	Significance
$1/n_F$	Related to the intrinsic heat of solute adsorption.
K_F	Equal to adsorption capacity when equilibrium concentration = 1.

*Adapted from Chiou, 2002. pp. 41 - 43.

1.7.0 Project overview

For the foreseeable future, the demand for products based on nanotechnology is expected to drastically increase (NSF, 2001). Carbon nanotubes and titanium dioxide nanoparticles are two types of nanomaterials which form the backbone of the nanotechnology industry. Applications for carbon nanotubes include high-strength material composites, semiconductors, and energy storage devices (Baughman et al., 2002). Titanium dioxide nanoparticles have been used in paints, photovoltaic cells, and chemical sensors (Varghese et al., 2003 and Reddya et al., 2002). Both nanoparticle titanium dioxide and carbon nanotubes have been found to be toxic to humans and animals (Brumfiel, 2003). In general, the toxicity of a carbon nanomaterial is not exclusively dependent upon its size, and may be augmented by toxins adsorbed to the nanoparticle (Colvin, 2003). Because of these toxicity risks, combined with the growing market for nanomaterials, it is of utmost importance that the impacts of nanomaterials on the environment be studied (Linkov et al., 2007, and Helland et al., 2008). Furthermore, investigating the interaction between known environmental toxins and nanomaterials is of utmost importance.

Antimony is used widely in flame retardants and automotive brake pads, but has also found increasing use in the semi-conductor industry (Van Velzen and Langenkamp, 1998). Recently, antimony-doped tin oxide films have been used in conjunction with carbon nanotubes to create transparent conducting films (Castro and Schmidt, 2007). Such films are not only desirable for their being both transparent and conductive, but also have the potential to be used as chemical sensors (Liu et al. 2008). The toxicity of antimony is well-documented (Chen et al. 2007, Newton et al. 1994, Oorts et al. 2008). The increasing ubiquity of manufactured nanomaterials merits the study of the interactions between antimony species and nanoparticles.

Carbon nanotubes are well-known for their adsorption of a variety of substances. Compared to the large number of studies of the sorption of metals or organic compounds by carbon nanotubes, relatively few studies have been completed regarding the sorption of metalloids by carbon nanotubes. As of 2008, no study has investigated the sorption capabilities between antimony and carbon nanotubes. The adsorption of a wide range of elements by titanium dioxide has been studied previously (Goncalves et al. 1999, Kajitvichyanukul et al. 2003, Thomas et al. 1998). Most of these studies do not specify the usage of nanoparticle titanium dioxide; and of those that do, the sorption of antimony by titanium dioxide nanoparticles has not been reported (Zhang, 2008). The sorption of ^{125}Sb on titanium dioxide surfaces has been studied (Mukhopadhyay and Lahiri, 2007). ^{125}Sb, however, is not a naturally occurring isotope. While the actual adsorption process between different isotopes of antimony is likely mechanistically identical, the adsorption of naturally-occurring isotopes of antimony to micro- and nanoparticle titanium dioxide still requires investigation.

In this study, we seek to investigate the antimony adsorption capability of multi-walled carbon nanotubes and titanium dioxide nanoparticles. The adsorption capacity of pristine MWCNTs will be compared to that of oxidized MWCNTs and activated carbon. Additionally, the adsorption capacity of nanoparticle titanium dioxide will be compared to that of microparticle titanium dioxide.

To investigate the adsorption affinity between the selected adsorbents and antimony (or other trace elements), a series of aqueous solutions containing antimony and/or trace elements will be prepared. To each solution will be added a known amount of adsorbent, and the solutions will be allowed to reach equilibrium. Inductively coupled plasma-atomic emission spectroscopy (ICP-AES) or inductively coupled plasma-mass spectrometry (ICP-MS) will be used to analyze the analyte concentrations in the solution fractions. The amount of antimony or trace elements adsorbed to the carbon nanotubes will be determined using the formula

$$q = \frac{V(C_i - C_{eq})}{m_a} \qquad (8)$$

where V is the total solution volume (L), C_i is the initial analyte concentration (mg/L), C_{eq} is the equilibrium analyte concentration (mg/L), m_a is the adsorbent mass (g), and q is the adsorbed concentration of analyte (mg/g). The data obtained will be used to generate adsorption isotherms, and be fit to linear, Freundlich and Langmuir isotherms using the isotherm fitting software IsoFit (SUNY Buffalo, Buffalo, NY, USA).

In the case of antimony adsorption by nanoparticle and microparticle titanium dioxide, the effects of light and adsorption on antimony speciation will be explored. High performance liquid chromatography (HPLC) using ion chromatography will be used in conjunction with ICP-MS to investigate inorganic antimony speciation, Sb(III) and Sb(V).

2.0 MATERIALS AND METHODS

The two nanomaterials investigated in this study were nanoparticle titanium dioxide (n-TiO2) and multi-walled carbon nanotubes (MWCNTs). To provide a comparison adsorbent for nanoparticle titanium dioxide, microparticle titanium dioxide was also utilized. Activated carbon was used as a control for comparison to MWCNTs and nanoparticle titanium dioxide, as it is a well-characterized adsorbent.

Table 2.0: Properties of materials used in experiments

Material	Supplier	Purity	Mean Particle Dimensions	Mean Surface Area (m^2g^{-1})	Zeta Potential (mV)
Activated Carbon (AC)	Fischer Scientific	99%	75 – 300 μm	500	N/A
Multi-walled Carbon Nanotubes (MWCNTs)	Chengdu Organic Chemistry Co. Ltd.	95%	Outer Diameter: 20 – 40 nm Length: 1 – 2 μm	107	N/A
Microparticle Titanium Dioxide (μ-TiO2)	Acros	99%	Diameter: 0.2 μm	N/A	N/A
Nanoparticle Titanium Dioxide HR3 (n-TiO2)	Mingri Nano-material Industrial Co. Ltd.	≥99%	Diameter: 50±5 nm	325	-30.0

2.1.0 Preparation of Sb (III) and Sb(V) solutions

One hundred (100.0 mL) milliliters of 1000 µg/mL antimony (III) solution were prepared by dissolving 0.2743 g of potassium antimonyl tartrate trihydrate ($C_8H_4K_2O_{12}Sb_2$ $3H_2O$, 0.411 mmol) with distilled water in a nitric acid leached volumetric flask wrapped in aluminum foil. The flask was stirred for 24 hours before being refrigerated at 4 °C. Antimony (V) solution (100.0 mL, 1000 µg/mL) was prepared by dissolving 0.2158 g potassium hexahydroxy antimonite ($KSb(OH)_6$) with distilled water in a volumetric flask. The flask was stirred for 48 hours before being refrigerated at 4 °C.

2.2.0 Procedure for the surface oxidation of multi-walled carbon nanotubes

Multi-walled carbon nanotubes were surface oxidized using a modification of a procedure published by Li et al. (2003). MWCNTs (200.0 mg) were submerged in 20.0 mL of 15.8 M nitric acid in a Teflon digester vessel. A Hot Block Pro Environmental Express digester (Mt. Pleasant, SC, USA) was used to heat the vessel to 100 °C for 2 hours. After removing the digesting tube from the digester, and allowing the tube to cool, the supernatant was decanted from the vessel. The oxidized MWCNTs were transferred to a centrifuge tube, and washed with distilled water. After centrifugation, the supernatant was decanted, and the washing process was repeated four more times. After pouring off the supernatant for the final time, the oxidized CNTs were transferred to an oven and allowed to dry for 48 hours at 80 °C. Final yield was 130 mg (65% w/w).

2.3.0 Leaching of multi-walled carbon nanotubes

In order to assess the presence of lead, antimony, and nickel already present in the MWCNTs, a leaching test was performed. A series of 12, 15-mL polypropylene centrifuge tubes were set up (see Table 2.3). To each of six tubes were added 25 mg of MWCNTs. Five (5.00 mL) of a 2% nitric acid solution was added to each of the six tubes; and for three of these tubes, concentrated (18 M) ammonium hydroxide was used to adjust the pH to 6.5. The remaining six tubes served as controls; all originally contained 5.00 mL 2% w/w nitric acid, but the pH of three of the controls was adjusted to 6.5. The tubes were all shaken for a period of seven days. The MWCNTs were removed from the solutions using 0.22 µm syringe filters, and the solutions were analyzed using a Perkin Elmer Elan 6000 ICP-MS (Perkin Elmer Instruments, Shelton, CT, USA) (see procedure 2.11.0 – "Procedure for analysis of samples by ICP-MS and ICP-AES").

Table 2.3: Leaching of multi-walled carbon nanotubes

Tube No.	Contents
1-3	25 mg MWCNTs + 5.00 mL 2% w/w nitric acid
4-6	5.00 mL 2% w/w nitric acid
7-9	25 mg MWCNTs + 5.00 mL 2% w/w nitric acid; pH adjusted to 6.5 with NH4OH
10-12	5.00 mL 2% w/w nitric acid; pH adjusted to 6.5 with NH4OH

2.4.0 Assessment of the effect of multi-walled carbon nanotube concentration on antimony adsorption at low pH

To each of six 50 mL polypropylene centrifuge tubes was introduced a mass of MWCNTs (20, 40, 50, 80, 150, and 200 mg). For comparison, six additional tubes were each filled with activated carbon (20, 40, 50, 80, 150, and 200 mg). To each tube was added 25 mL of a 26 µg/mL antimony (V) solution (pH: 1.38) diluted from a 1000 µg/mL standard prep solution from Spex Centriprep Standards (Metuchen, NJ). A control tube was set up containing only antimony solution. The tubes were shaken for 5 days at room temperature, and then allowed to settle overnight. Ten milliliters (10 mL) of the supernatant from each tube was removed using a precision pipette, and placed into a respective 15 mL centrifuge tube. The series of tubes was then centrifuged at 3000 rpm for 10 minutes. Following the centrifugation, 5 mL of solution was removed from each centrifuge tube and diluted with 5 mL distilled water. The solutions were then analyzed for antimony using a Perkin Elmer Optima 2000DV ICP-AES (Perkin Elmer Instruments, Shelton, CT, USA) (see procedure 2.11.0 – "Procedure for analysis of samples by ICP-MS and ICP-AES").

2.5.0 Investigating the effect of multi-walled carbon nanotube surface oxidation on Sb, Cd, and Pb adsorption capacity

Table 2.5: Experimental Setup

Control Tube No.	MWCNT Tube No.	Oxidized- MWCNT Tube No.	Conc. of Analyte Solution (μg/mL)
1-c, 2-c	1-m, 2-m	1-o, 2-o	0.05
3-c, 4-c	3-m, 4-m	3-o, 4-o	0.10
5-c, 6-c	5-m, 6-m	5-o, 6-o	0.50
7-c, 8-c	7-m, 8-m	7-o, 8-o	1.00
9-c, 10-c	9-m, 10-m	9-o, 10-o	5.00
11-c, 12-c	11-m, 12-m	11-o, 12-o	10.00
13-c, 14-c	13-m, 14-m	13-o, 14-o	20.00

Using the method described in "Procedure for the surface oxidation of multi-walled carbon nanotubes," (see section 2.2.0) MWCNTs (200 mg) were surface oxidized and prepared for use in the experiment. Fourteen polypropylene centrifuge tubes were each filled with 10 mg of pristine MWCNTs. Fourteen more tubes were filled with 10 mg surface oxidized MWCNTs. A remaining fourteen tubes were left empty (to be filled with control solutions). All 42 tubes were then filled; two from each group were filled with 10 mL of each concentration gradient (see table 2.5 above). The antimony, lead, and cadmium concentrations used were 50 ng/mL, 100 ng/mL, 500 ng/mL, 1 μg/mL, 5 μg/mL, 10 μg/mL, and 20 μg/mL.

Following a 1-week equilibration period, samples were filtered through 0.22 μm Teflon syringe filters and analyzed by either a Perkin Elmer Elan 6000 ICP-MS (Perkin Elmer Instruments, Shelton, CT, USA) (for analyte concentrations of 50 and 100 ng/mL) or a Perkin Elmer Optima 2000DV ICP-AES (Perkin Elmer Instruments, Shelton, CT, USA) (for concentrations of 0.5 – 20 μg/mL) (see procedure 2.11.0 – "Procedure for analysis of samples by ICP-MS and ICP-AES").

2.6.0 Determination of the effect of analyte (Pb, Sb, and Cd) concentration on the adsorption potential of MWCNTs, activated carbon (AC), micro- and nanoparticle titania

For each of four adsorbents (MWCNTs, AC, microparticle titania or nanoparticle titania), twelve 15 mL polypropylene centrifuge tubes were each filled with 20 mg of the respective adsorbent. For each adsorbent, each set of two of the 12 tubes was filled with 10 mL of a solution containing lead (II), antimony (V), and cadmium (II) (made by diluting solution standards from Spex Centriprep Standards (Metuchen, NJ)).

The pH of the solutions used had been previously adjusted to approximately 6.3 using 18M ammonium hydroxide. All tubes were shaken for one week at room temperature, and then allowed to settle overnight. Ten milliliters (10 mL) of the supernatant from each tube was removed using a precision pipette, and filtered through a 0.22 μm syringe filter. The solutions were then analyzed using a Perkin Elmer Optima 2000DV ICP-AES (Perkin Elmer Instruments, Shelton, CT, USA) (see procedure 2.11.0 – "Procedure for analysis of samples by ICP-MS and ICP-AES").

Table 2.6: Determination of the effect of analyte (Pb, Sb, and Cd) concentration on the adsorption potential of MWCNTs, activated carbon (AC), micro- and nanoparticle titania (Experimental Setup)

Tube No.	Adsorbent	Solution Added (10.00 mL)
1, 2	MWCNT	0 μg/mL Cd (II), Pb (II), Sb (V)
3, 4	MWCNT	1 μg/mL Cd (II), Pb (II), Sb (V)
5, 6	MWCNT	5 μg/mL Cd (II), Pb (II), Sb (V)
7, 8	MWCNT	10 μg/mL Cd (II), Pb (II), Sb (V)
9, 10	MWCNT	20 μg/mL Cd (II), Pb (II), Sb (V)
11, 12	MWCNT	50 μg/mL Cd (II), Pb (II), Sb (V)
13, 14	AC	0 μg/mL Cd (II), Pb (II), Sb (V)
15, 16	AC	1 μg/mL Cd (II), Pb (II), Sb (V)
17, 18	AC	5 μg/mL Cd (II), Pb (II), Sb (V)
19, 20	AC	10 μg/mL Cd (II), Pb (II), Sb (V)
21, 22	AC	20 μg/mL Cd (II), Pb (II), Sb (V)
23, 24	AC	50 μg/mL Cd (II), Pb (II), Sb (V)
25, 26	μ-TiO$_2$	0 μg/mL Cd (II), Pb (II), Sb (V)
27, 28	μ-TiO$_2$	1 μg/mL Cd (II), Pb (II), Sb (V)
29, 30	μ-TiO$_2$	5 μg/mL Cd (II), Pb (II), Sb (V)
31, 32	μ-TiO$_2$	10 μg/mL Cd (II), Pb (II), Sb (V)
33, 34	μ-TiO$_2$	20 μg/mL Cd (II), Pb (II), Sb (V)
35, 36	μ-TiO$_2$	50 μg/mL Cd (II), Pb (II), Sb (V)
37, 38	n-TiO$_2$	0 μg/mL Cd (II), Pb (II), Sb (V)
39, 40	n-TiO$_2$	1 μg/mL Cd (II), Pb (II), Sb (V)
41, 42	n-TiO$_2$	5 μg/mL Cd (II), Pb (II), Sb (V)
43, 44	n-TiO$_2$	10 μg/mL Cd (II), Pb (II), Sb (V)
45, 46	n-TiO$_2$	20 μg/mL Cd (II), Pb (II), Sb (V)
47, 48	n-TiO$_2$	50 μg/mL Cd (II), Pb (II), Sb (V)

2.7.0 Investigating the antimony adsorption capacity of nanoparticle titanium dioxide

To each of a series of fourteen 15-mL polypropylene centrifuge tubes was added 20 mg of nanoparticle titanium dioxide. Solutions of antimony (V) were prepared in a series of concentrations (50 ng/mL, 100 ng/mL, 500 ng/mL, 1 μg/mL, 5 μg/mL, 10 μg/mL, 20 μg/mL), and for each concentration, two centrifuge tubes were each filled with 10 mL of the solution (pH ~3.9). All tubes were shaken for a period of 48 hours, at which point the solutions were filtered through 0.22 μm Teflon filters and analyzed by either a Perkin Elmer Elan 6000 ICP-MS (Perkin Elmer Instruments, Shelton, CT, USA) (for analyte concentrations of 50 and 100 ng/mL) or a Perkin Elmer Optima 2000DV ICP-AES (for analyte concentrations of 0.5 – 20 µg/mL) (Perkin Elmer Instruments, Shelton, CT, USA) (see procedure 2.11.0 – "Procedure for analysis of samples by ICP-MS and ICP-AES").

2.8.0 Investigating Sb(V) adsorption by micro (µ-TiO_2)- and nanoparticle titanium dioxide (n-TiO_2); effects on speciation

The difference in Sb(V) adsorption capacity of micro- and Nanoparticle titanium was explored. A series of six 15-mL polypropylene centrifuge tubes was set up. Two of the tubes were filled with 20 mg of nanoparticle titanium dioxide, two more were filled with 20 mg of microparticle titanium dioxide. The remaining two tubes were left empty. All tubes were then filled with 10 mL of 50 ng/mL Sb(V) (500 ng of Sb) solution. Final pH was 3.93. All tubes were shaken and allowed to equilibrate for a period of 48 hours. Following the equilibration period, the samples were filtered through a 0.22µm Teflon filter, and analyzed using a Beckman Gold HPLC (Beckman Coulter Inc., Fullerton, CA, USA) coupled to a Perkin Elmer Elan 6000 ICP-MS (Perkin Elmer Instruments, Shelton, CT, USA) (IC-ICP-MS), using the method developed by Steely et al., (2007) (see procedure 2.12.0 – "Procedure for the analysis of solution samples by HPLC-ICP-MS").

2.9.0 Investigating the effect of light upon antimony (III) adsorption by nano- and microparticle titanium dioxide

The effect of light on the adsorption of Sb(III) was investigated. A series of twelve 15-mL polypropylene centrifuge tubes was set up. Four of the tubes were filled with 20 mg of nanoparticle titania, four more were filled with 20 mg of microparticle titania. The remaining four tubes were left empty (as controls). All tubes were then filled with 10 mL of 50 ng/mL Sb(III) solution. Two tubes from each set of four had been covered in aluminum foil to protect the contents from light. All tubes were shaken and allowed to equilibrate for a period of 48 hours. Following the equilibration period, the samples were filtered through a 0.22 µ m Teflon filter, and analyzed using a Beckman Gold HPLC (Beckman Coulter Inc., Fullerton, CA, USA) coupled to a Perkin Elmer Elan 6000 ICP-MS (Perkin Elmer Instruments, Shelton, CT, USA) (IC-ICP-MS), using the method developed by Steely et al., (2007) (see procedure 2.12.0 – "Procedure for the analysis of solution samples by HPLC-ICP-MS"). Duplicate analyses were performed.

2.10.0 Obtaining infrared spectra of n-TiO2 and Sb adsorbed on n-TiO$_2$

Small portions (~10 mg) of pristine nanoparticle titanium dioxide and nanoparticle titanium dioxide containing adsorbed antimony (approximately 10 µg Sb/mg n-TiO$_2$) were dried in an oven at 80°C for a period of 48 hours. Samples were then mixed with potassium bromide in a ratio of 2 mg sample per 98 mg KBr, and ground into a fine power. Infrared spectra were obtained by diffuse reflectance infrared Fourier transform (DRIFT) spectroscopy, using a Midac series M-2000 Fourier transformed infrared spectrophotometer (Irvine, CA, USA) equipped with a DRIFT accessory (Spectros Instruments, Shrewsbury, MA, USA).

2.11.0 Procedure for the analysis of solution samples by ICP-AES and ICP-MS

Elemental concentration data were obtained using a Perkin Elmer Optima 2000DV ICP-AES (Shelton, CT, USA) or a Perkin Elmer Elan 6000 ICP-MS (Perkin Elmer Instruments, Shelton, CT, USA). To create calibration curves, standards were prepared from Spex Centriprep Standards (Metuchen, NJ). The experimental parameters for sample analysis by ICP-MS and ICP-AES are shown in Tables 2.11a and 2.11b respectively.

Table 2.11a: ICP-AES Operational Setup

Spectrometer Model:	Optima 2000 DV
Autosampler Model:	AS-91
Nebulizer Back Pressure (kPa):	300.0 - 320.0
Nebulizer Flow Rate (L/min):	ICP Continuous
Technique:	0.80
Analyte Wavelength (nm):	Cd(I) 228.802, Ni(II) 231.604, Pb(II) 220.353, Sb(II) 206.836

Table 2.11b: ICP-MS Operational Setup

Model:	Perkin Elmer Elan 6000
Forward RF Power (W):	1000
Argon Gas Flows (L/min):	15 (coolant), 0.92 (auxiliary), 1.40 (carrier)
Nebulizer:	Cross Flow (Perkin Elmer)
Spray Chamber:	Scott-Type (Ryton)
Dwell Time (ms):	1000 (chromatography), 50 (quantitative analysis)
Sweeps/Reading:	1 (chromatography), 20 (quantitative analysis)
Readings/Replicate:	115 (chromatography), 1 (quantitative analysis)
Replicates:	1 (chromatography), 6 (quantitative analysis)
Isotopes Monitored:	^{121}Sb, ^{123}Sb, and 220—Background mass, ^{209}Bi, ^{72}Ge (internal standards used for quantitative analysis)

2.12.0 Procedure for the analysis of solution samples by HPLC-ICP-MS

For antimony speciation work, samples were analyzed using a Beckman Gold HPLC (Beckman Coulter Inc., Fullerton, CA, USA) interfaced with a Perkin Elmer Elan 6000 ICP-MS (Perkin Elmer Instruments, Shelton, CT, USA). Experimental setup parameters are shown in Table 2.12.

Table 2.12: HPLC Operational Setup

Model:	Beckman Gold (Beckman Coulter Inc., Fullerton, CA, USA)
Column:	PRP X-100 Anion Exchange (Hamilton, Reno, NV, USA)
Guard Column:	Hamilton PRP 100 guard column
Mobile Phase:	20 mM Na_2EDTA and 2 mM KHP, pH 4.0
Pump Pressure (kpsi):	1.0 - 1.1
Flow Rate (mL/min):	1.0
Elution Mode:	Isocratic
Injection Volume (µL):	100
Run Time (s):	~400

2.13.0 Fitting adsorption data to linear, Langmuir, and Freundlich isotherms

For all adsorption experiments, equilibrium concentration and adsorption data were manipulated using Microsoft Excel (Redmond, WA, USA). Isotherm fits of the data were calculated using IsoFit (SUNY Buffalo, Buffalo, NY, USA). Fits were determined by using a combination of particle swarm optimization and the non-linear Levenberg-Marquardt regression algorithm (Matott, 2007). Once the isotherm parameters were calculated, graphical representations of the data and respective isotherm fits were plotted in Excel.

3.0 RESULTS AND DISCUSSION

3.1.0 Leaching of MWCNTs

In order to assess the presence of Pb, Sb, and Ni already present in multi-walled carbon nanotubes, a leaching experiment was set up. Two blank solutions containing 2% w/w nitric acid were prepared; the pH of one of the solutions was adjusted to 7.0 using concentrated ammonium hydroxide. MWCNTs were added to two additional nitric acid solutions. All samples were then allowed a week to equilibrate. Following the equilibration period, MWCNTs were removed by filtration using a 0.22 μm syringe filter, and the remaining solution was analyzed using ICP-MS.

Levels of lead and antimony in the leachate solutions were insignificant as measured by ICP-MS. As expected, there was a significant amount of nickel remaining in the carbon nanotubes (see Fig. 3.1a). The reason for this observation is likely due to the fact that the original synthesis was by chemical vapor deposition (CVD) using a nickel catalyst. It has been previously determined that particles from the original metal catalyst often remain embedded in CNTs synthesized by CVD (Yuan et al., 2001).

As a result of the nickel already present in the MWCNTs, a correction factor had to be applied to any experiment involving nickel adsorption. Without the correction factor, results appeared as shown in Figure 3.1b.

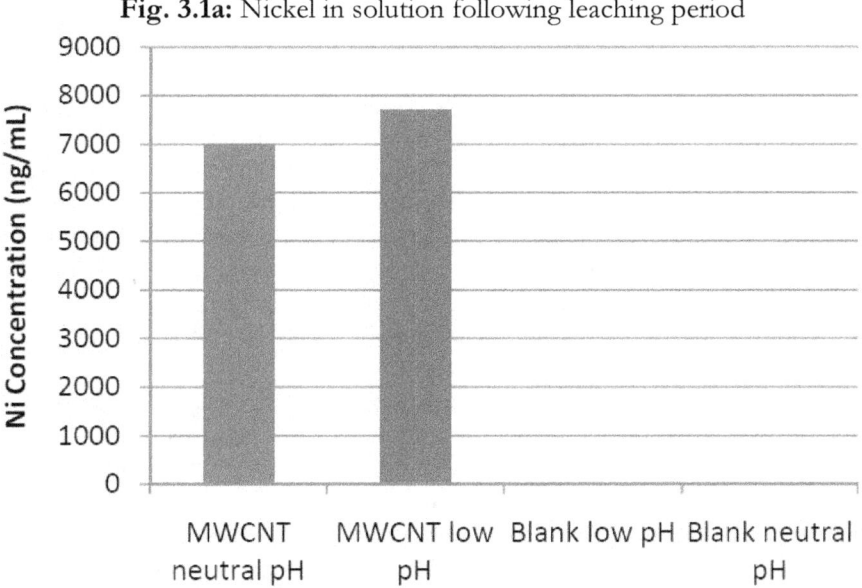

Fig. 3.1a: Nickel in solution following leaching period

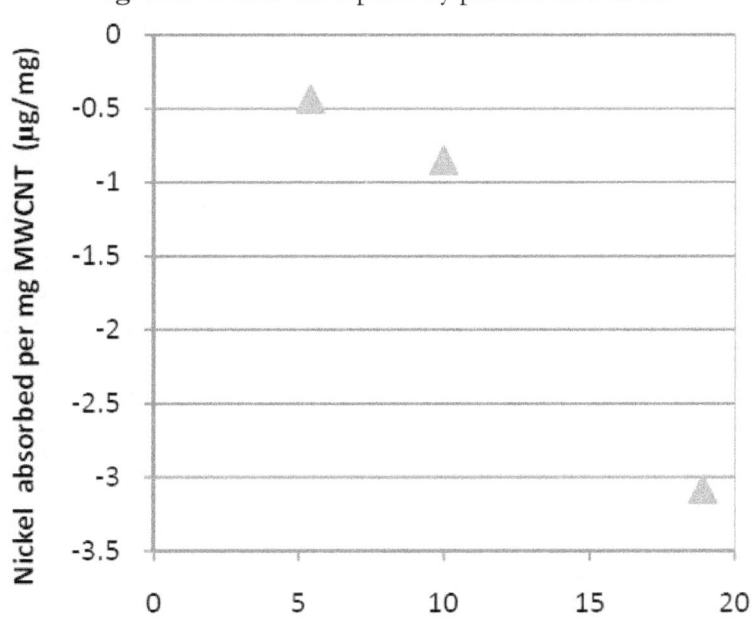

Fig. 3.1b: Nickle adsorption by pristine MWCNTs

3.2.0 Assessing the relationship between antimony adsorption capacity and multi-walled carbon nanotube concentration at low pH

To assess the dependence of antimony adsorption capacity of pristine multi-walled carbon nanotubes on MWCNT concentration, a series of solutions containing antimony (20 µg/mL) were prepared; to each solution was added a different amount of pristine MWCNTs to create heterogeneous mixtures of MWCNT concentrations 0.8 mg/mL, 1.6 mg/mL, 2 mg/mL, 3.2 mg/mL, 4 mg/mL, and 6 mg/mL. The pH was measured (~1.38), but left unmodified. A replicate experiment was also set up, using AC in place of MWCNTs. After an equilibration period of 5 days, samples (i.e. remaining liquid phase) were analyzed by ICP-AES.

As can be seen in figure 3.2, the antimony adsorption by the MWCNTs is negligible when compared to the antimony adsorption exhibited by AC. The correlation coefficient ($R^2 = 0.114$) for the MWCNT adsorption data suggests that the variation displayed by the data points is nearly random in nature. This supports the negligible antimony adsorption displayed by the MWCNTs at low pH, which was expected. The correlation coefficient ($R^2 = 0.869$) of the AC adsorption data, however, is moderately strong. The data clearly indicate that with increased AC concentration, a greater fraction of the total antimony is adsorbed.

Fig. 3.2: Fraction of total antimony adsorption as a function of MWCNT or AC concentration

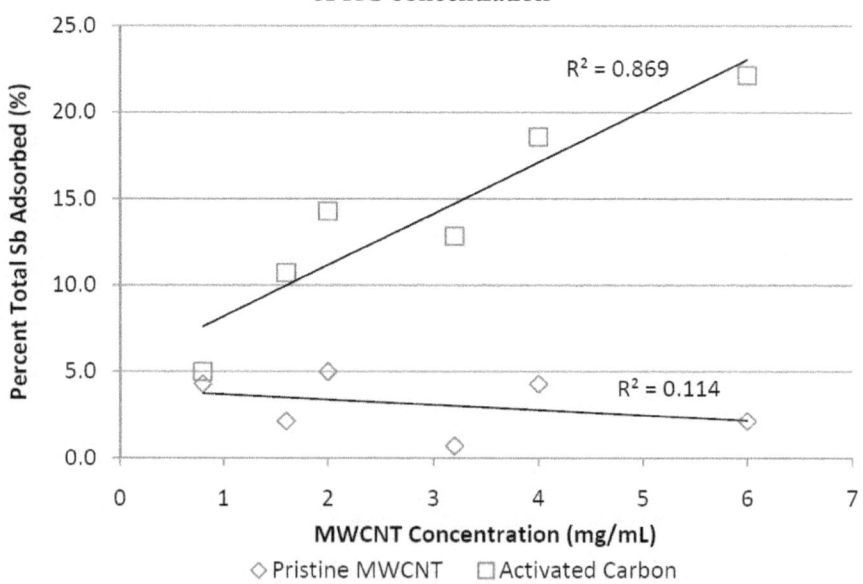

3.3.0 Investigating the effect of MWCNT surface oxidation on Sb (V), Cd (II), and Pb (II) adsorption capacity

Following the surface oxidation of MWCNTs using nitric acid, an experiment was performed to compare the antimony, cadmium, and lead adsorption capacity of pristine MWCNTs and surface-functionalized MWCNTs. A wide range of analyte (adsorbate) concentrations was used (0.05, 0.1, 0.5, 1, 5, 10, and 20 μg/mL); the pH of the equilibration solutions was left unmodified (pH ~ 1.4). Following the equilibration period (1 week), the trace element concentrations in the solutions were measured by ICP-MS and ICP-AES.

As is seen in figures 3.3a and 3.3b, a simple plot of the concentration of adsorbed analyte (Cd (II), Pb (II), or Sb(V)) versus the initial concentration of the respective analyte, results in a fairly linear plot. As the initial analyte concentration is increased, the concentration of the adsorbed analyte also increases. The relationship appears linear, and if a linear regression is performed, correlation coefficients range from 0.9593 – 0.9967 (see Table 3.3a).

Fitting the adsorption data to the Langmuir isotherm resulted in acceptable correlation coefficient values (R^2 = 0.9599 – 0.9967). Freundlich isotherm fits of the data resulted in similar correlation coefficient values (R^2 = 0.9074 – 0.9947). Extracted Langmuir and Freundlich parameters are displayed in Tables 3.3b and 3.3c, respectively. In general, a greater K_L and $1/n_F$ implies a greater attraction between adsorbent and adsorbate, while the magnitude of q_m is related to the adsorption capacity of an adsorbent for a given adsorbate. The Langmuir results of this experiment suggest that the strongest adsorbent-adsorbate interaction is between that of Pb and pristine MWCNTs, while the Freundlich results demonstrate that the interactions are approximately equal in strength between all adsorbent-adsorbate pairs studied. The monolayer adsorption capacity for antimony is the highest in all cases studied.

It appears that at the pH studied, surface-functionalized MWCNTs possess a nearly identical adsorption profile of antimony, cadmium, and lead (when compared to pristine MWCNTs). The adsorption of the analytes at pH ~ 1.4 of both the pristine MWCNTs and the surface oxidized MWCNTs is miniscule compared to the adsorption capacity of MWCNTs at a more neutral pH (~6.4, where the analyte adsorption by MWCNTs is roughly 3 times greater). It is therefore suspected that, had the pH of the equilibration solutions been higher, increased adsorption of all analytes would have been observed. Additionally, it is possible that the difference in adsorption capacity between pristine MWCNTs and surface-functionalized MWCNTs would have been greater. Comparing the individual adsorption characteristics of the three analytes, neither pristine MWCNTs nor surface oxidized MWCNTs appeared to indicate a preference for adsorption of antimony, cadmium, or lead. It is suspected that this is a result of the low level of analyte adsorption, since the solution pH (~1.4) is apparently too low for any significant adsorption to occur.

Fig. 3.3a: Linear isotherm of the MWCNT adsorption of antimony, lead and cadmium. pH = 1.4.

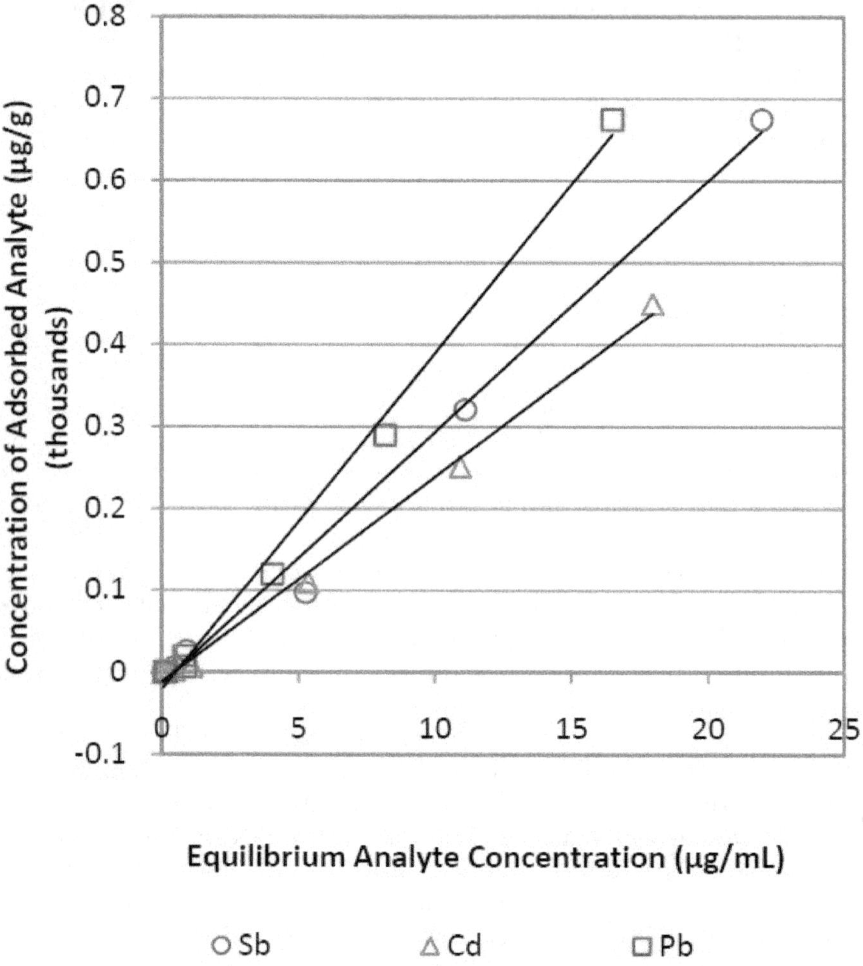

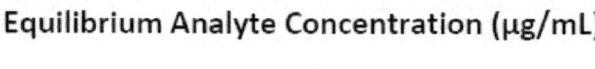

Fig. 3.3b: Linear isotherm of the surface-oxidized MWCNT adsorption of antimony, lead and cadmium. pH = 1.4.

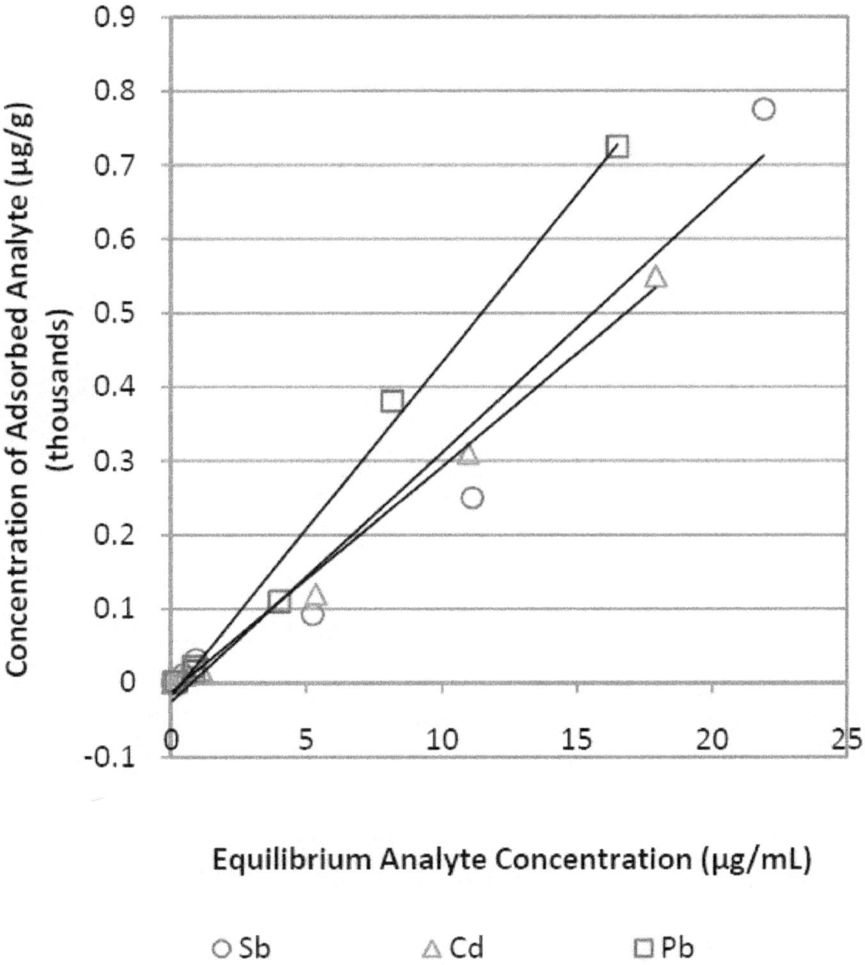

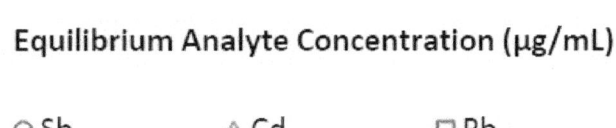

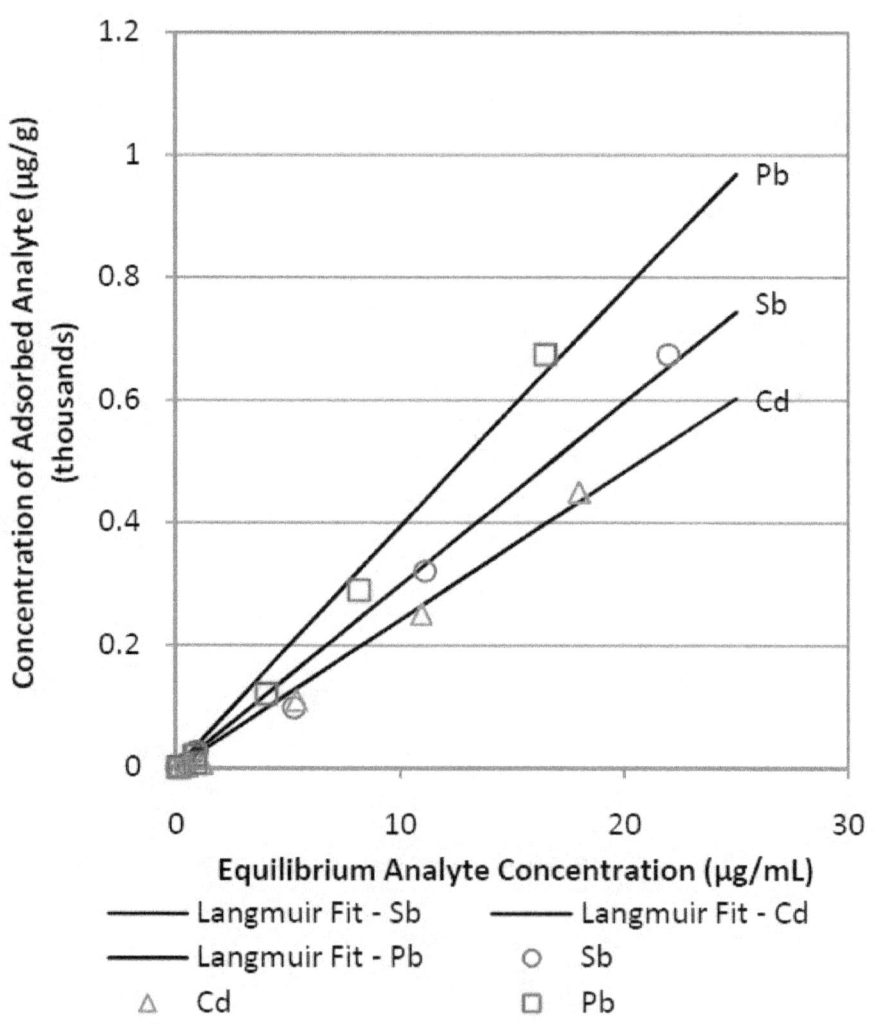

Fig. 3.3c: Langmuir isotherm of the MWCNT adsorption of antimony, lead and cadmium. pH = 1.4.

Fig. 3.3d: Langmuir isotherm of the surface-oxidized MWCNT adsorption of antimony, lead and cadmium. pH = 1.4.

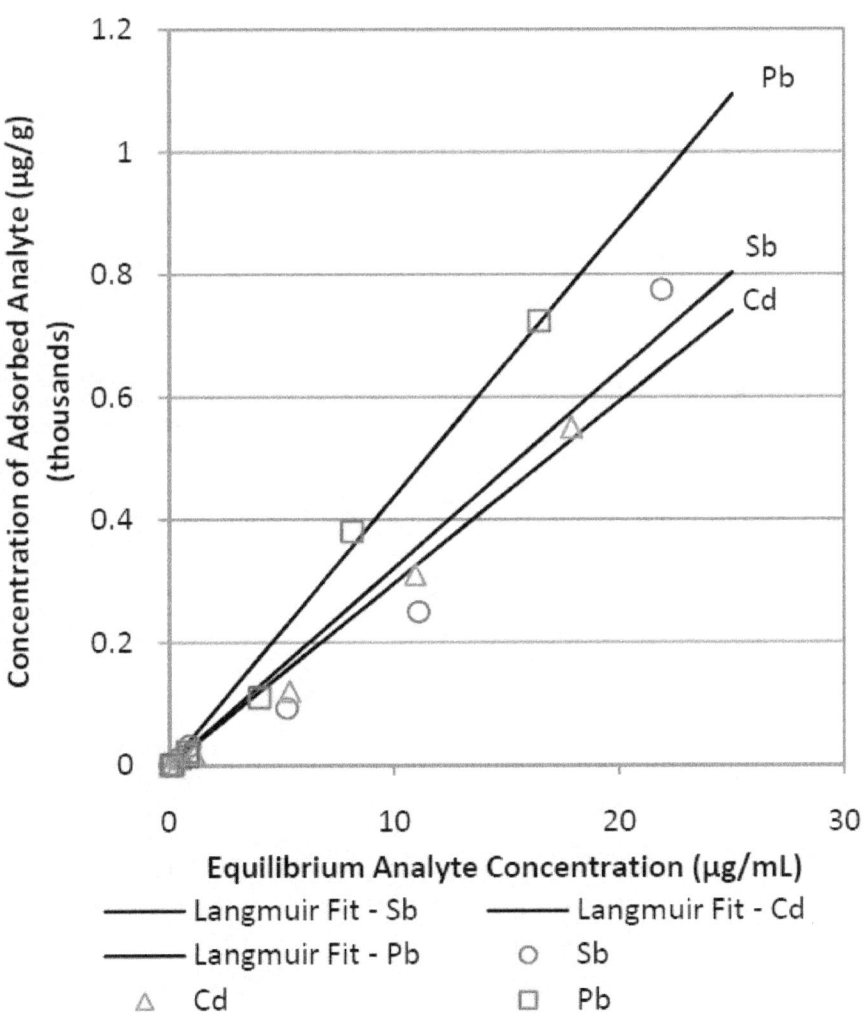

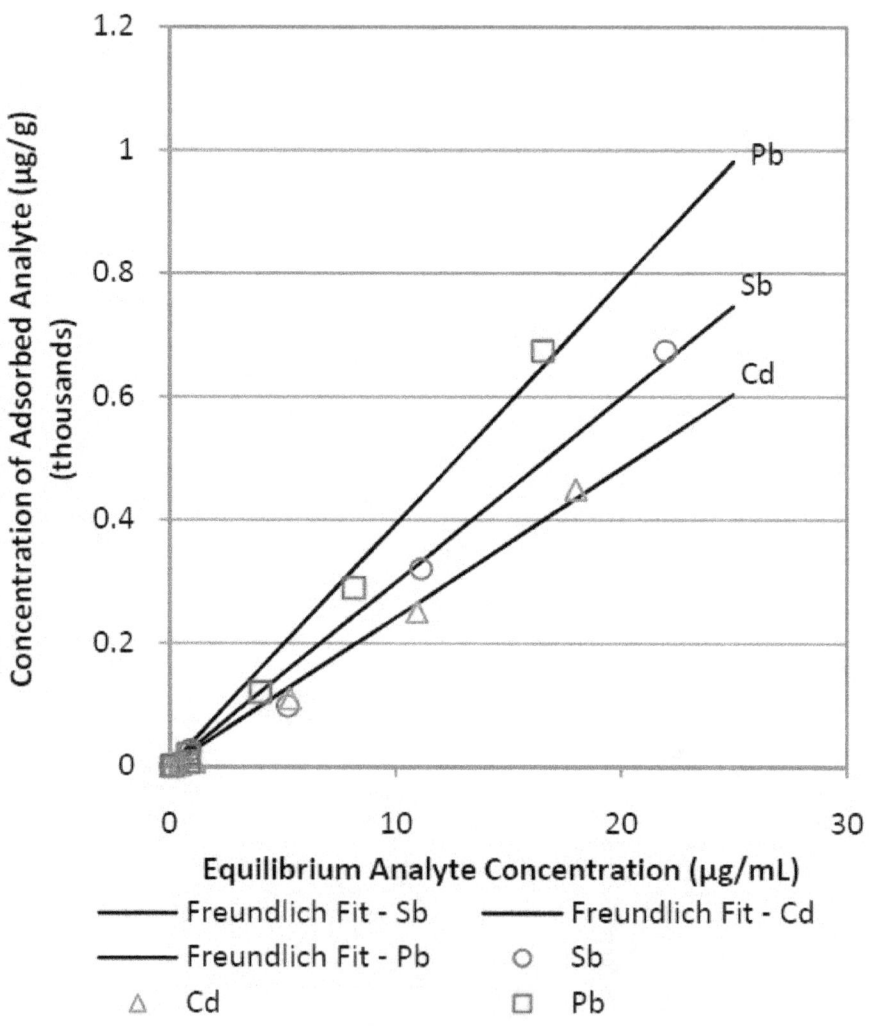

Fig. 3.3e: Freundlich isotherm of the MWCNT adsorption of antimony, lead and cadmium. pH = 1.4.

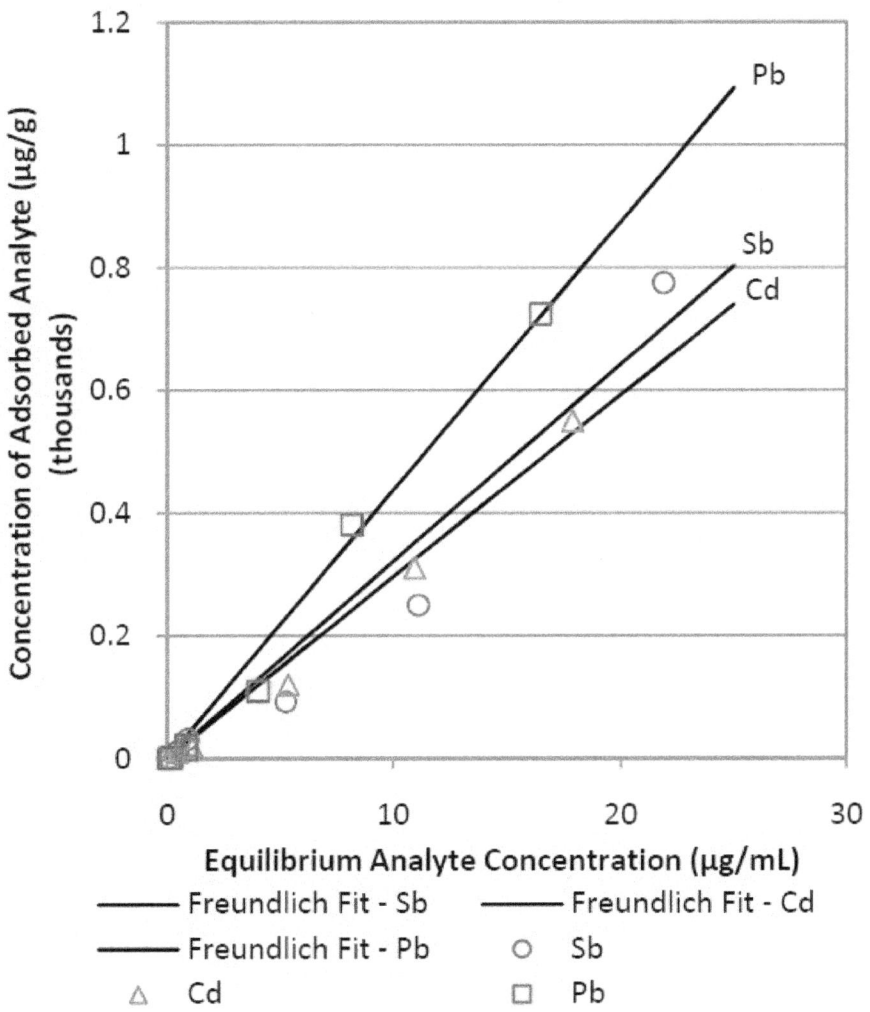

Fig. 3.3f: Freundlich isotherm of the surface-oxidized MWCNT adsorption of antimony, lead and cadmium. pH = 1.4.

Table 3.3a: Linear Regression Parameters

Material	Correlation Coefficient	Slope	Y-intercept
Pristine MWCNT - Sb	0.9913	0.030	-0.013
Pristine MWCNT - Cd	0.9967	0.025	-0.011
Pristine MWCNT - Pb	0.9948	0.041	-0.021
Oxidized MWCNT - Sb	0.9593	0.033	-0.025
Oxidized MWCNT - Cd	0.9938	0.030	-0.012
Oxidized MWCNT - Pb	0.9909	0.045	-0.018

Table 3.3b: Determined Langmuir Parameters

Material	Correlation Coefficient	q_m	K_L
Pristine MWCNT - Sb	0.9926	1.504×10^3	1.984×10^{-5}
Pristine MWCNT - Cd	0.9967	1.434×10^3	1.684×10^{-5}
Pristine MWCNT - Pb	0.9948	3.889×10^1	1.023×10^{-3}
Oxidized MWCNT - Sb	0.9599	1.977×10^3	1.625×10^{-5}
Oxidized MWCNT - Cd	0.9935	7.762×10^1	3.830×10^{-4}
Oxidized MWCNT - Pb	0.9917	5.303×10^1	8.334×10^{-4}

Table 3.3c: Determined Freundlich Parameters

Material	Correlation Coefficient	$1/n_F$	K_F
Pristine MWCNT - Sb	0.9924	1.000×10^{-1}	2.989×10^{-2}
Pristine MWCNT - Cd	0.9947	9.999×10^{-1}	3.927×10^{-2}
Pristine MWCNT - Pb	0.9074	9.998×10^{-1}	2.417×10^{-2}
Oxidized MWCNT - Sb	0.9599	9.999×10^{-1}	3.214×10^{-2}
Oxidized MWCNT - Cd	0.9947	9.995×10^{-1}	2.967×10^{-2}
Oxidized MWCNT - Pb	0.9919	9.998×10^{-1}	4.374×10^{-2}

3.4.0 Determination of the effect of analyte (Pb, Sb, and Cd) concentration on the adsorption potential of MWCNTs, activated carbon (AC), micro- and nanoparticle titanium dioxide (μ-TiO$_2$ and n-TiO$_2$)

An experiment was performed to compare the antimony, cadmium, and lead adsorption capacity of pristine multi-walled carbon nanotubes, activated carbon, microparticle titanium dioxide, and nanoparticle titanium dioxide at near neutral pH. A range of analyte concentrations was used (1, 5, 10, 20 and 50 µg/mL), and the pH of the equilibration solutions was increased from ~ 1.4 to ~ 6.4 using ammonium hydroxide (18M). Following the equilibration period (1 week), results were collected by ICP-AES.

As is seen in Figs. 3.4a – 3.4d, a simple plot of the concentration of adsorbed analyte (Cd, Pb, or Sb) versus the equilibrium solution concentration of the respective analyte, results in a fairly linear plot. As the initial analyte concentration is increased, the concentration of the adsorbed analyte also increases. The relationship appears roughly linear, and if a linear regression is performed, correlation coefficients range from 0.8134 – 0.9893 (see Table 3.4a).

Considering the simple adsorption plots (see Figs. 3.4a – 3.4d), several trends are apparent. First, nanoparticle titanium dioxide has the greatest adsorption capacity for all of the analytes studied (Sb, Cd, Pb), followed by activated carbon, microparticle titanium dioxide, and pristine MWCNTs (MWCNT <μ-TiO$_2$ < AC < n-TiO$_2$). In general, among all adsorbents, lead tends to be the analyte most easily adsorbed. Interestingly, nanoparticle titanium dioxide exhibits significant adsorption of antimony (Fig. 3.4d) when compared to the remaining three adsorbents (AC, μ-TiO$_2$ and n-TiO$_2$), which all adsorb antimony to approximately the same degree (Figs. 3.4a – 3.4c). Comparing MWCNTs to the activated carbon control, activated carbon is a far superior adsorbent for lead, and a slightly better adsorbent for cadmium (Figs. 3.4a and 3.4b). In the case of antimony, the adsorption capability of AC and MWCNTs is nearly identical. Finally, making a direct comparison between μ-TiO$_2$ and n-TiO$_2$ indicates that nanoparticle titanium dioxide is a better adsorbent for all analytes when compared to microparticle titanium dioxide (Figs. 3.4c and 3.4d).

The Langmuir isotherm worked acceptably to describe analyte adsorption by all adsorbents studied (AC, MWCNT, μ-TiO$_2$, and n-TiO$_2$; for all analytes R^2 = 0.8748 – 0.9907), but provided marginal results when describing cadmium adsorption by MWCNTs and microparticle titanium dioxide (R^2 = 0.8728 – 0.8748) (Figs. 3.4e – 3.4h). The results of fitting the data to the Freundlich isotherm tended to mirror those of the Langmuir isotherm (Figs. 3.4i – 3.4l). Analyte adsorption by nanoparticle titanium dioxide was described the best (Fig. 3.4l; R^2 = 0.9568–0.9907), while cadmium adsorption by microparticle titanium dioxide and MWCNTs was described poorly (Figs. 3.4j and 3.4k; R^2 = 0.8746 and 0.8734, respectively). Of all analytes investigated, lead was most easily fit to both the Langmuir and Freundlich isotherms (Figs. 3.4i – 3.4l; R^2 = 0.9537 – 0.9907). Linear, Langmuir and Freundlich parameters were extracted from all of the plots, and are summarized in tables 3.4a - 3.4c. If a comparison of Freundlich and Langmuir parameters is made to those published for the interaction of functionalized MWCNTs with lead (Li et al., 2002), it is found that the values of q_m and K_F are larger, K_L is smaller, and n_F is similar in magnitude.

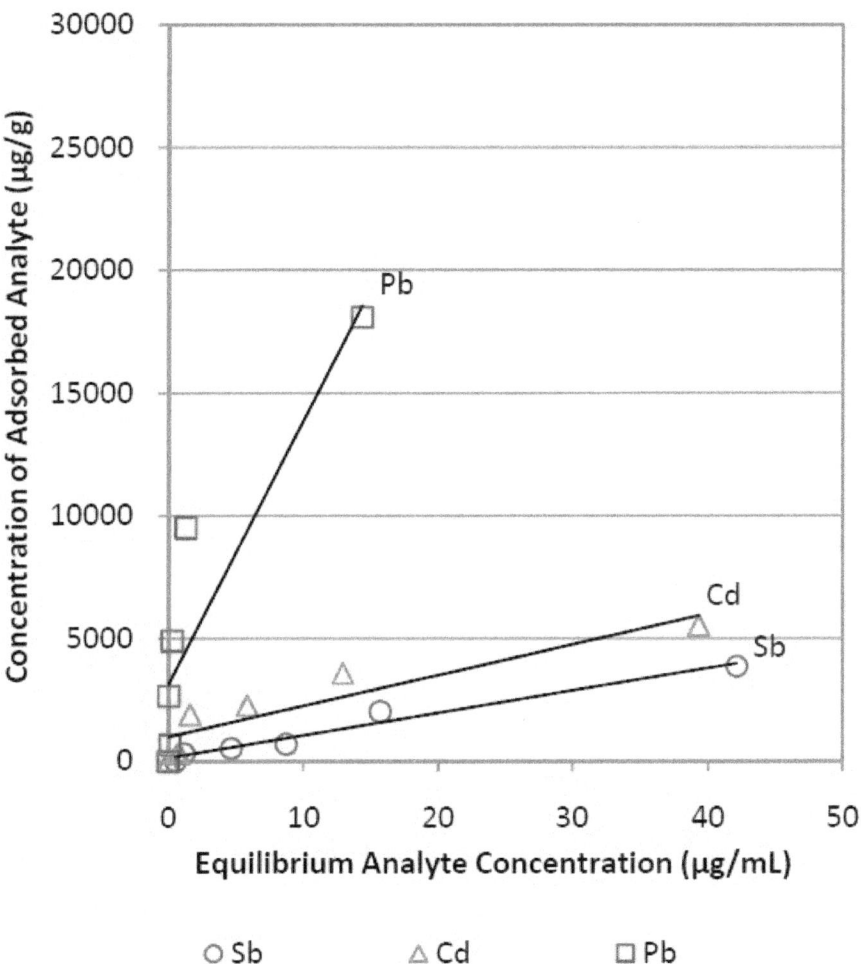

Fig. 3.4a: Antimony, cadmium and lead linear adsorption isotherms for activated carbon. pH = 6.4, equilibration period = 7 d.

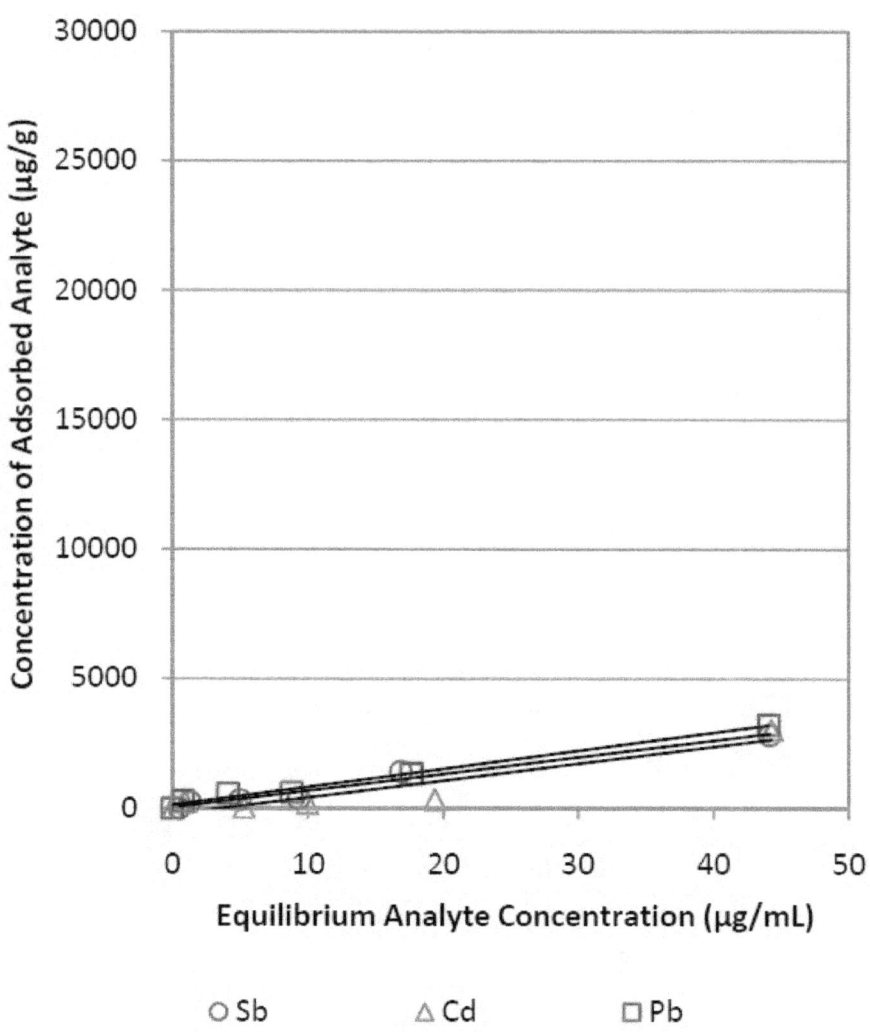

Fig. 3.4b: Antimony, cadmium and lead linear adsorption isotherms for pristine MWCNTs. pH = 6.4, equilibration period = 7 d.

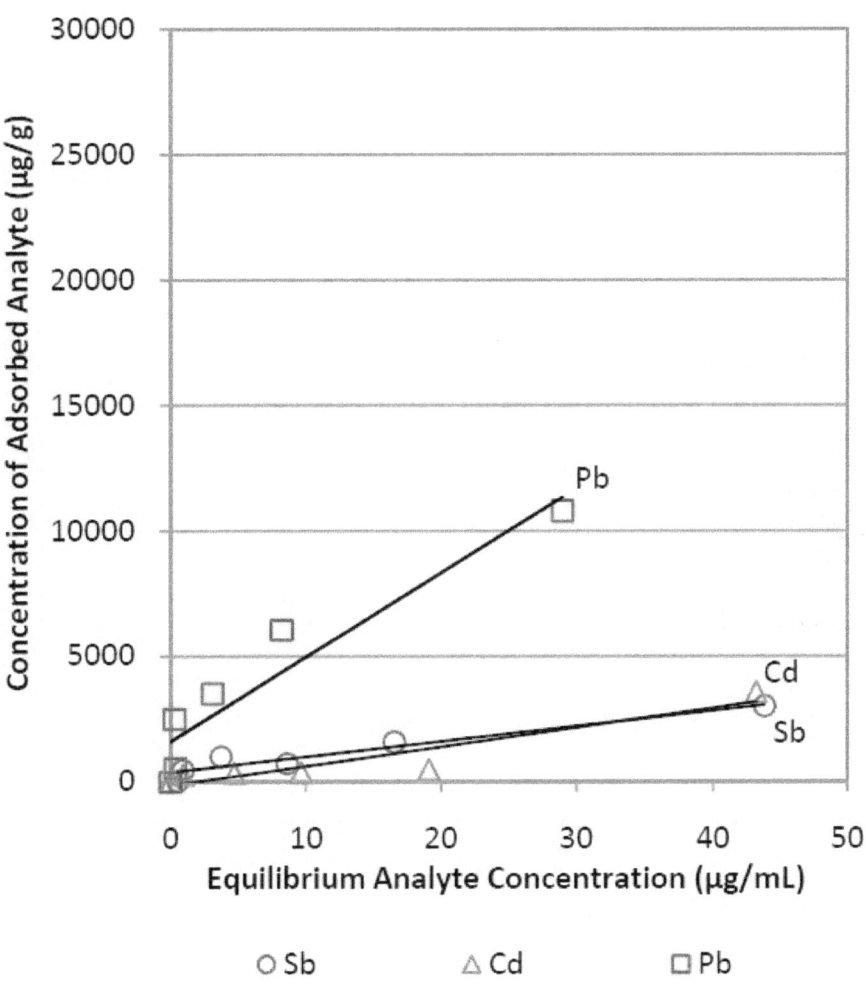

Fig. 3.4c: Antimony, cadmium and lead linear adsorption isotherms for microparticle titanium dioxide. pH = 6.4, equilibration period = 7 d.

Fig. 3.4d: Antimony, cadmium and lead linear adsorption isotherms for nanoparticle titanium dioxide. pH = 6.4, equilibration period = 7 d.

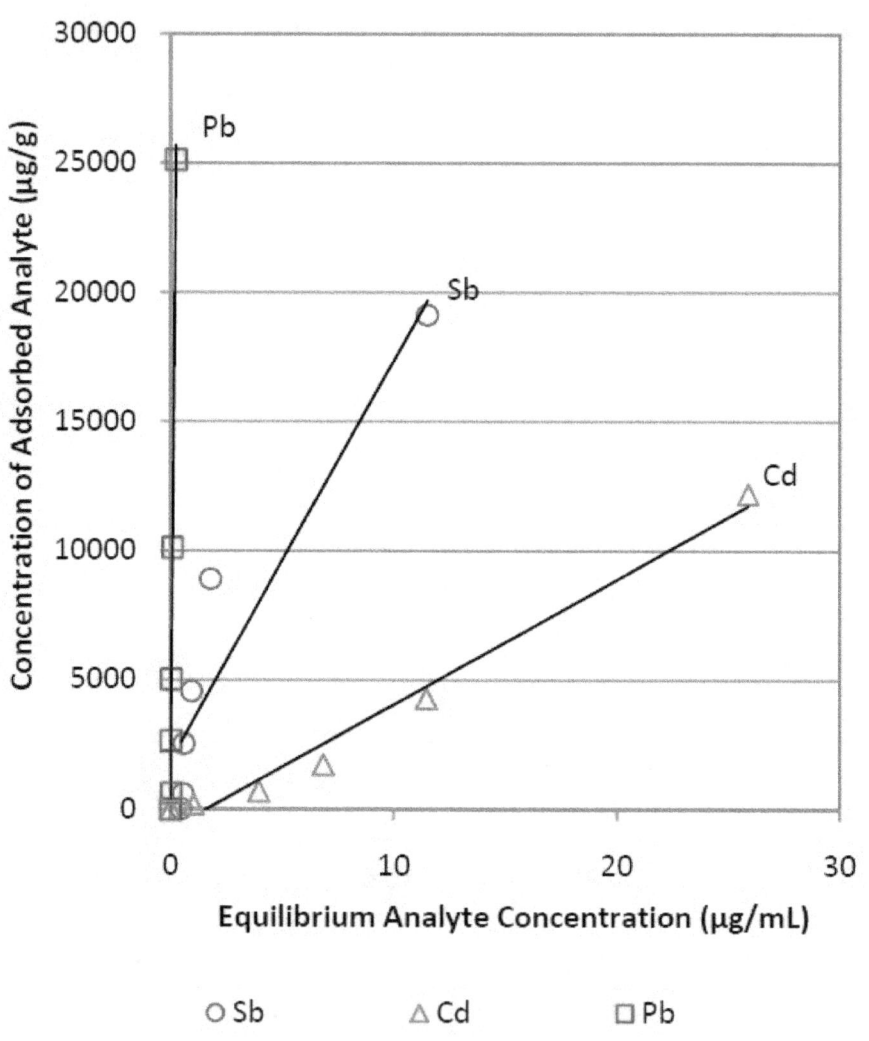

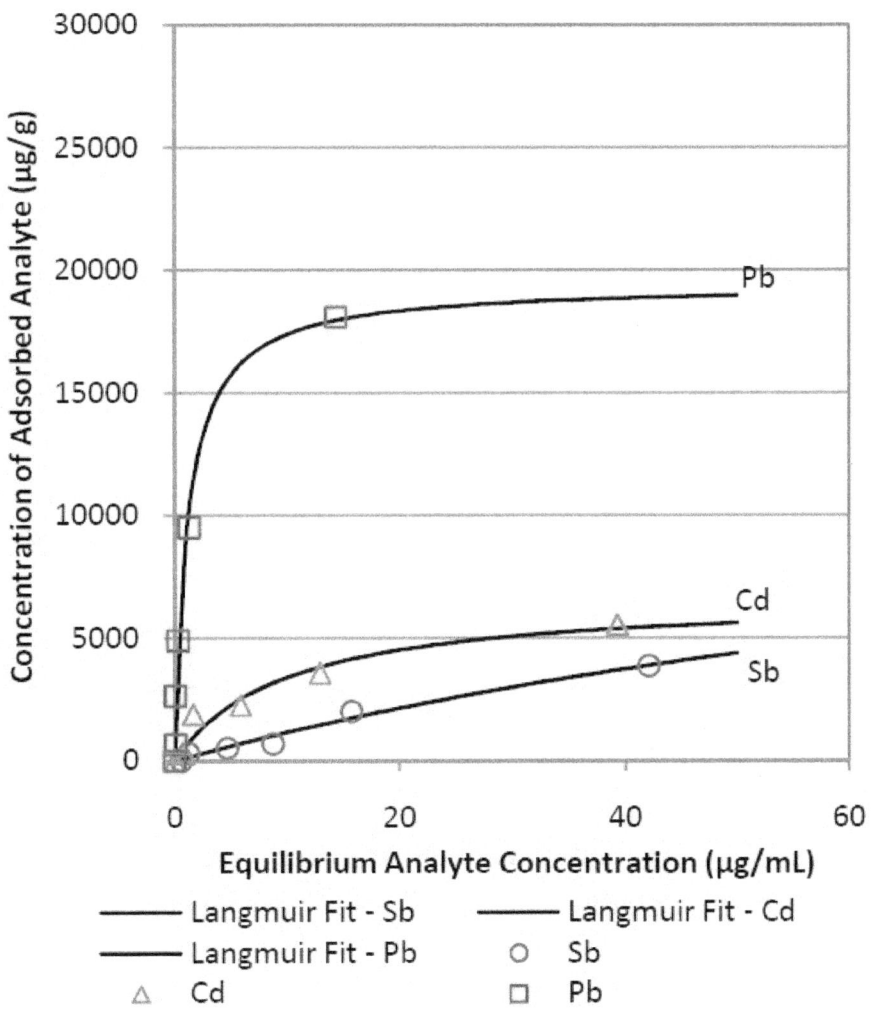

Fig. 3.4e: Antimony, cadmium and lead Langmuir adsorption isotherms for activated carbon. pH = 6.4, equilibration period = 7 d.

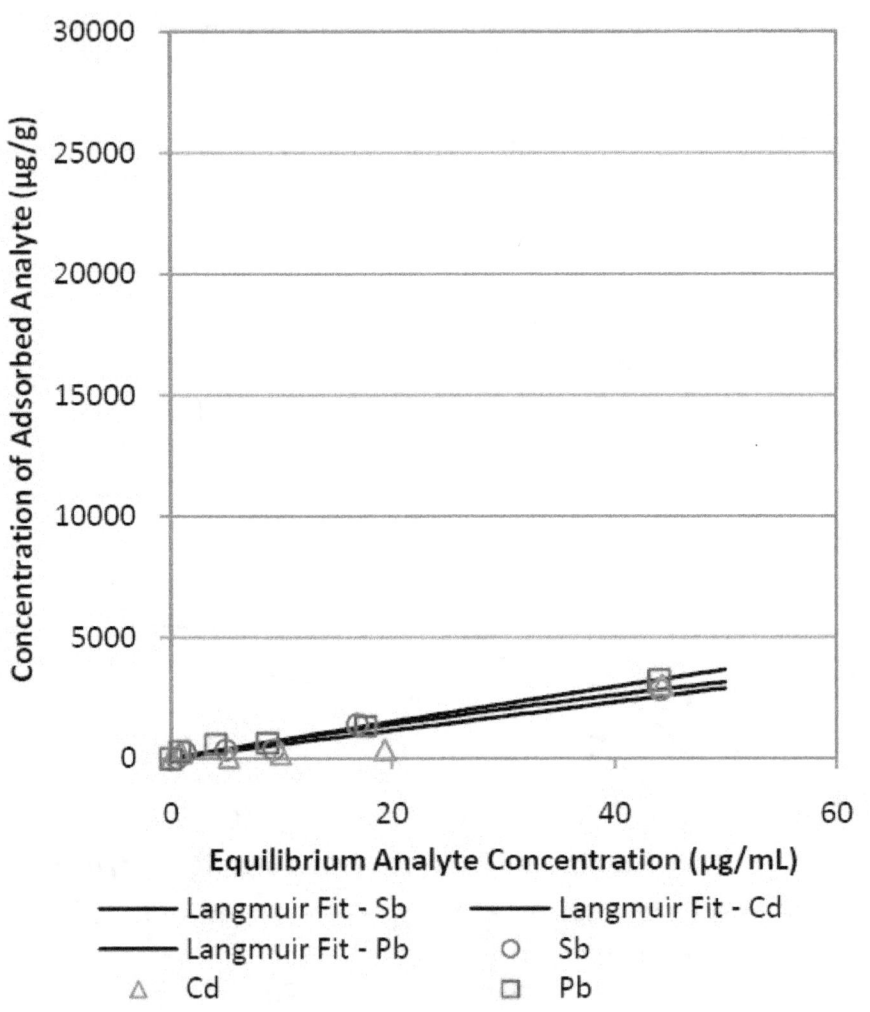

Fig. 3.4f: Antimony, cadmium and lead Langmuir adsorption isotherms for pristine MWCNTs. pH = 6.4, equilibration period = 7 d.

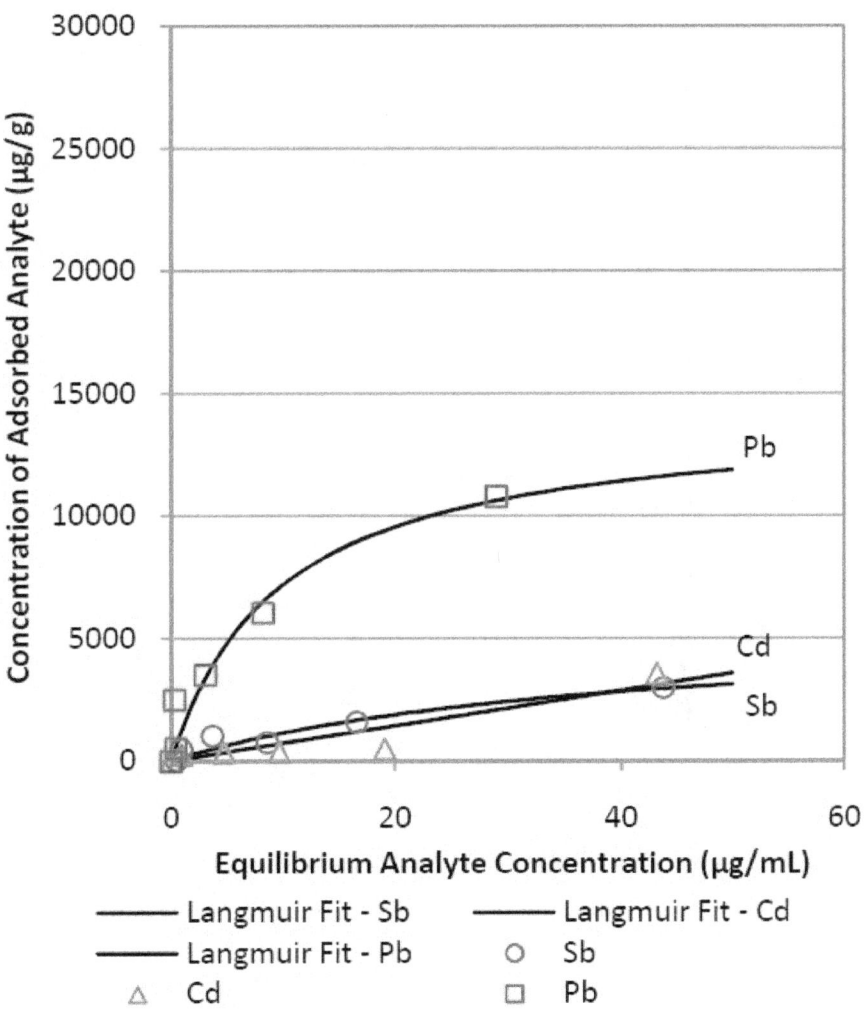

Fig. 3.4g: Antimony, cadmium and lead Langmuir adsorption isotherms for microparticle titanium dioxide. pH = 6.4, equilibration period = 7 d.

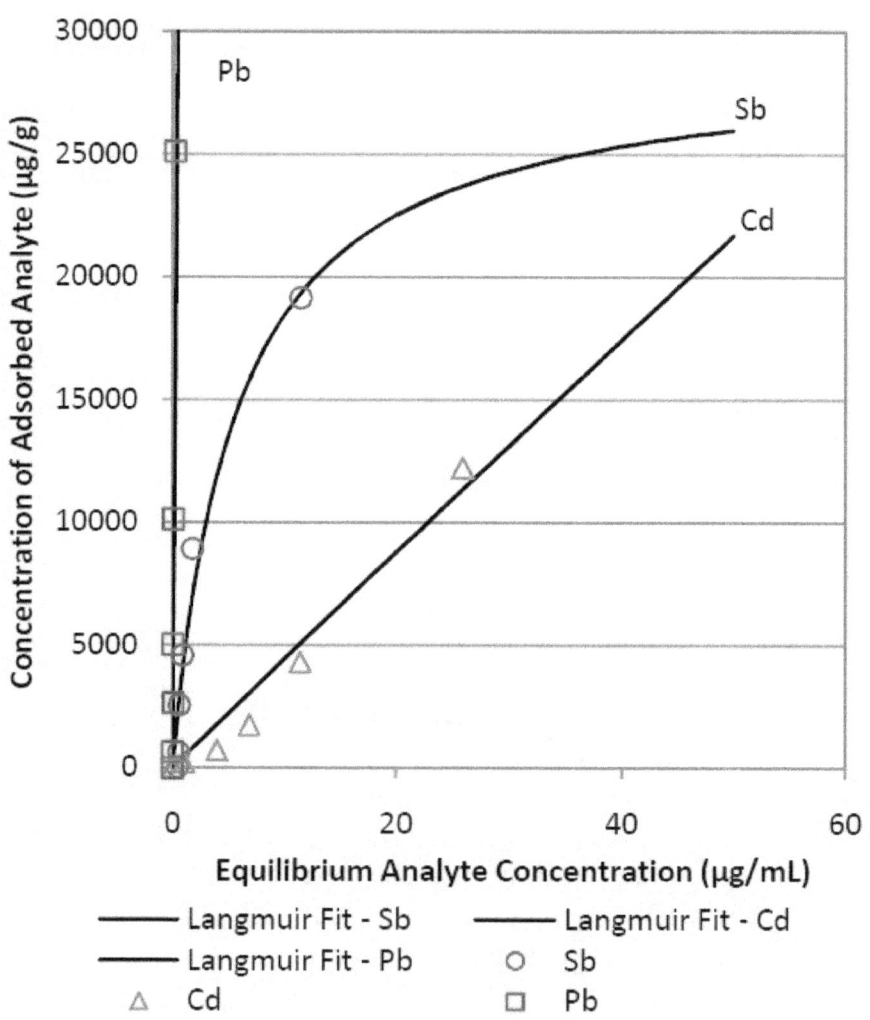

Fig. 3.4h: Antimony, cadmium and lead Langmuir adsorption isotherms for nanoparticle titanium dioxide. pH = 6.4, equilibration period = 7 d.

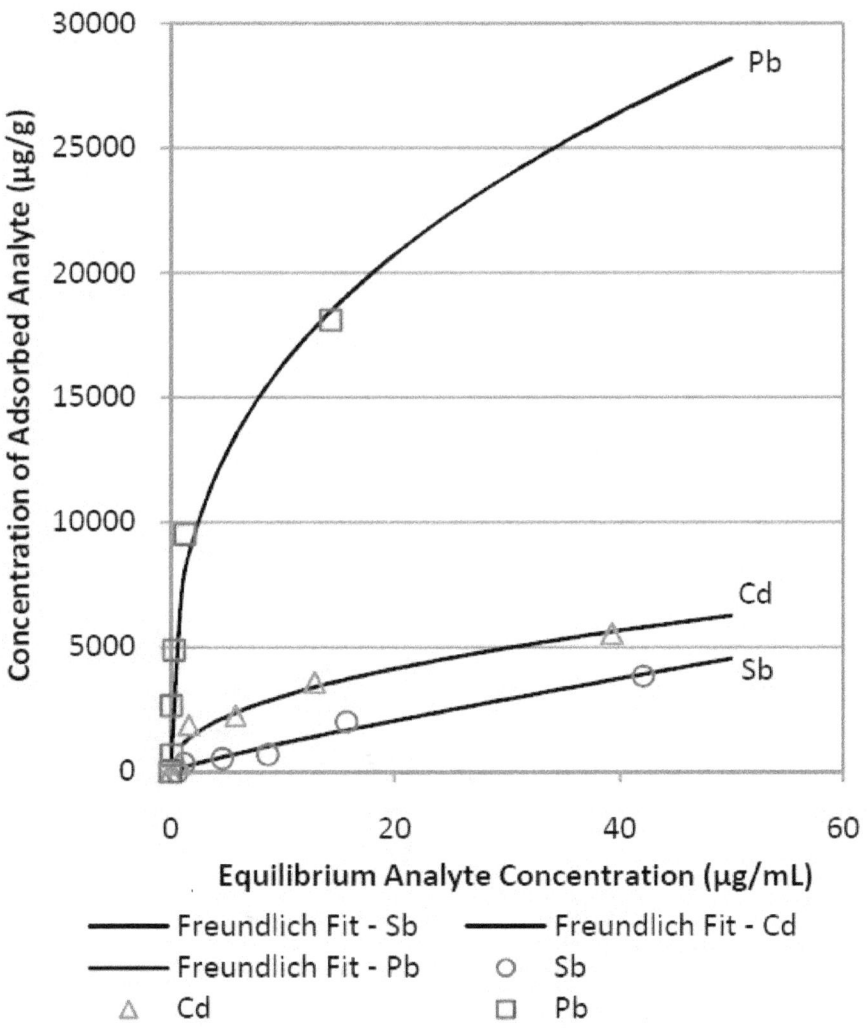

Fig. 3.4i: Antimony, cadmium and lead Freundlich adsorption isotherms for activated carbon. pH = 6.4, equilibration period = 7 d.

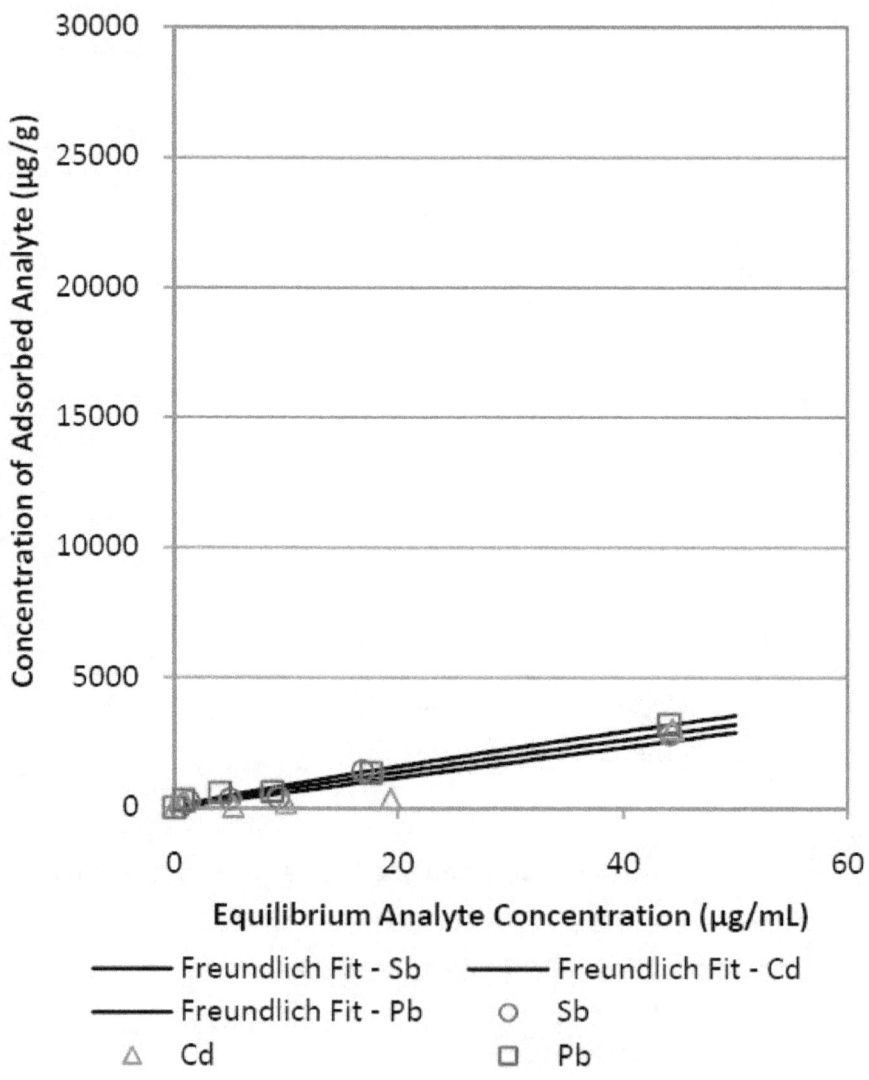

Fig. 3.4j: Antimony, cadmium and lead Freundlich adsorption isotherms for pristine MWCNTs. pH = 6.4, equilibration period = 7 d.

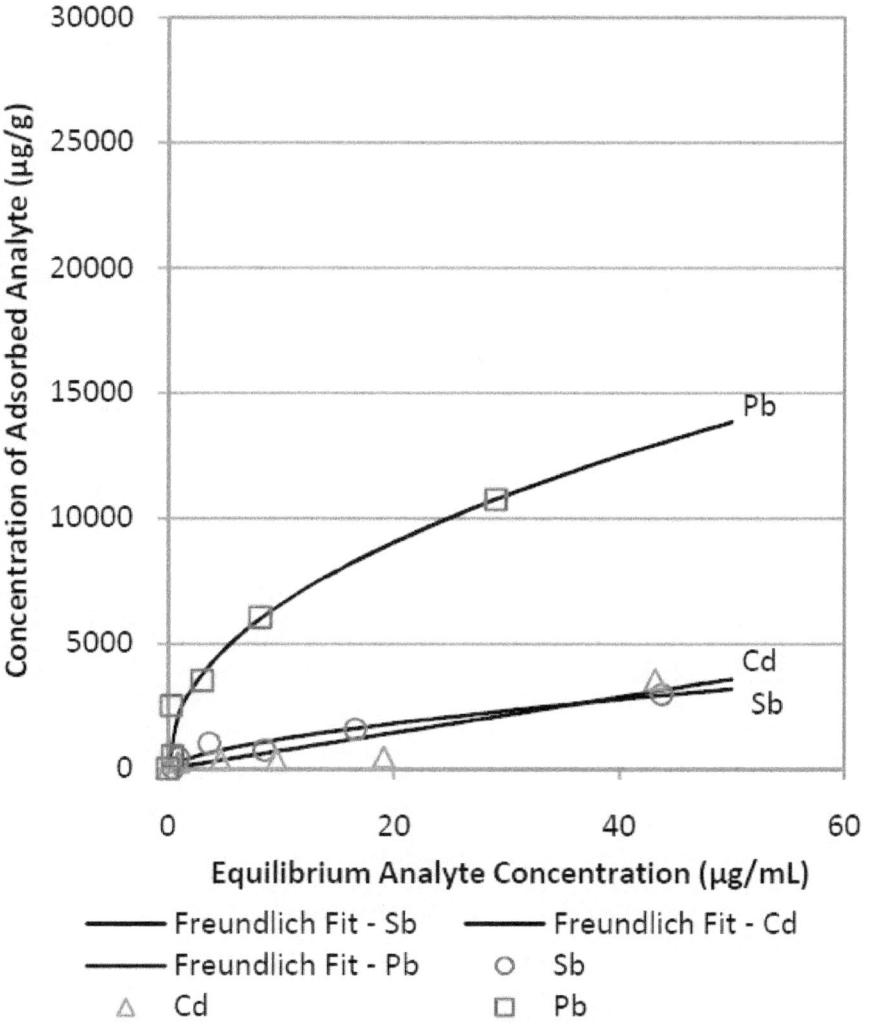

Fig. 3.4k: Antimony, cadmium and lead Freundlich adsorption isotherms for microparticle titanium dioxide. pH = 6.4, equilibration period = 7 d.

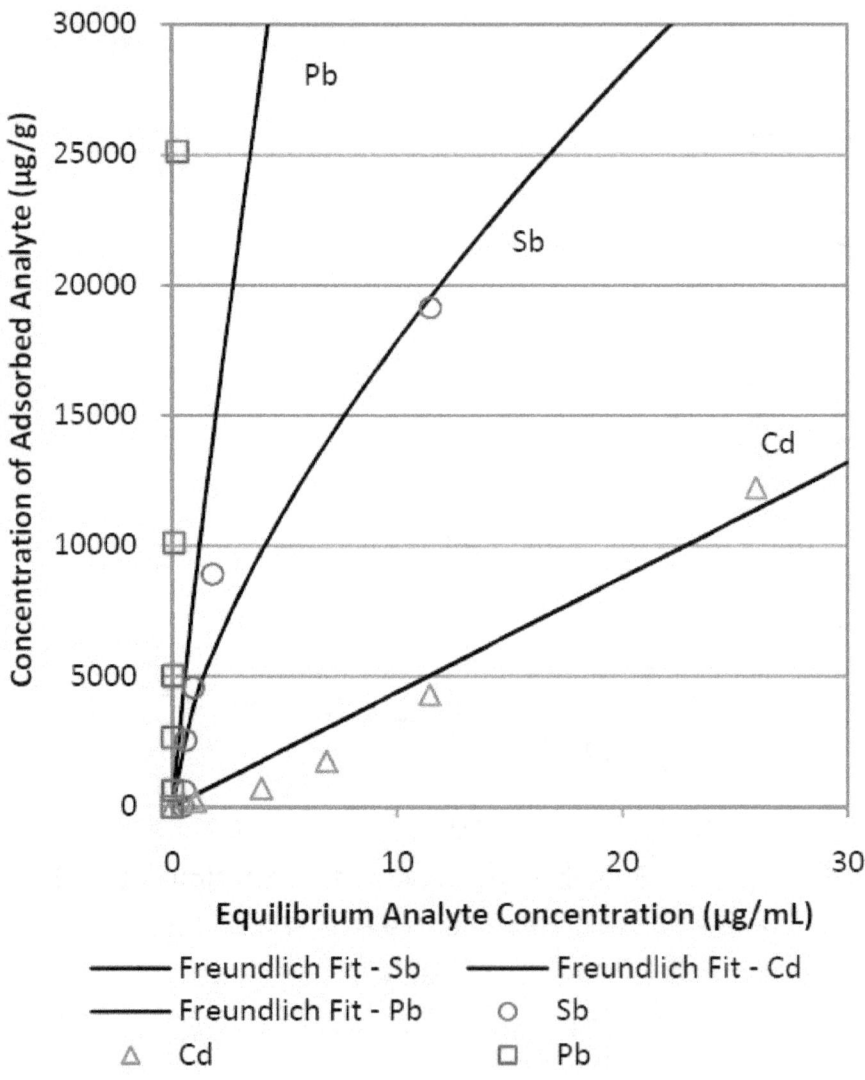

Fig. 3.4l: Antimony, cadmium and lead Freundlich adsorption isotherms for nanoparticle titanium dioxide. pH = 6.4, equilibration period = 7 d.

Table 3.4a: Linear Regression Parameters

Material	Correlation Coefficient	Slope	Y-intercept
Activated Carbon - Sb	0.9704	91.32	120.1
Activated Carbon - Cd	0.8402	125.6	991.3
Activated Carbon - Pb	0.8134	1081.0	3084.0
Pristine MWCNT - Sb	0.9695	64.16	49.86
Pristine MWCNT - Cd	0.8706	65.39	-226.8
Pristine MWCNT - Pb	0.9893	69.63	143.6
Microparticle titanium dioxide - Sb	0.9305	61.52	372.9
Microparticle titanium dioxide - Cd	0.8819	77.43	-168.3
Microparticle titanium dioxide - Pb	0.8938	335.8	1606.0
Nanoparticle titanium dioxide - Sb	0.8793	1552.0	1842.0
Nanoparticle titanium dioxide - Cd	0.9806	483.5	-771.6
Nanoparticle titanium dioxide - Pb	0.9838	94847	408.5

Table 3.4b: Determined Langmuir Parameters

Material	Correlation Coefficient	q_m	K_L
Activated Carbon - Sb	0.9799	1.452×10^4	8.656×10^{-3}
Activated Carbon - Cd	0.9387	6.692×10^3	1.032×10^{-1}
Activated Carbon - Pb	0.9739	1.942×10^4	8.523×10^{-1}
Pristine MWCNT - Sb	0.9724	1.757×10^4	4.380×10^{-3}
Pristine MWCNT - Cd	0.8728	9.099×10^5	6.932×10^{-5}
Pristine MWCNT - Pb	0.9895	5.900×10^4	1.325×10^{-3}
Microparticle titanium dioxide - Sb	0.9297	5.554×10^3	2.591×10^{-2}
Microparticle titanium dioxide - Cd	0.8748	8.427×10^5	8.552×10^{-5}
Microparticle titanium dioxide - Pb	0.9537	1.417×10^4	1.032×10^{-1}
Nanoparticle titanium dioxide - Sb	0.9568	2.896×10^4	1.751×10^{-1}
Nanoparticle titanium dioxide - Cd	0.9875	9.999×10^5	4.437×10^{-4}
Nanoparticle titanium dioxide - Pb	0.9907	1.054×10^5	1.179

Table 3.5c: Determined Freundlich Parameters

Material	Correlation Coefficient	$1/n_F$	K_F
Activated Carbon - Sb	0.9768	8.614×10^{-1}	1.556×10^2
Activated Carbon - Cd	0.9503	4.512×10^{-1}	1.069×10^3
Activated Carbon - Pb	0.9529	3.501×10^{-1}	7.264×10^3
Pristine MWCNT - Sb	0.9708	9.244×10^{-1}	8.626×10^1
Pristine MWCNT - Cd	0.8734	1.000	5.804×10^1
Pristine MWCNT - Pb	0.9876	8.772×10^{-1}	1.152×10^2
Microparticle titanium dioxide - Sb	0.9427	6.196×10^{-1}	2.831×10^2
Microparticle titanium dioxide - Cd	0.8746	1.000	7.188×10^1
Microparticle titanium dioxide - Pb	0.9629	4.656×10^{-1}	2.243×10^3
Nanoparticle titanium dioxide - Sb	0.9227	6.538×10^{-1}	3.958×10^3
Nanoparticle titanium dioxide - Cd	0.9879	1.000	4.396×10^2
Nanoparticle titanium dioxide - Pb	0.9876	8.894×10^{-1}	8.215×10^4

3.5.0 Determination of the antimony adsorption capacity of nanoparticle titanium dioxide

For a set concentration of nanoparticle titanium dioxide (2 mg/mL), a range of antimony solutions was prepared (0.05, 0.1, 0.5, 1, 5, 10, and 20 µg/mL) to investigate the antimony adsorption capacity of nanoparticle titanium dioxide. Across the ranges studied, it was determined that the adsorption was nearly complete, although at higher concentrations some antimony remained in solution (see Fig. 3.5a).

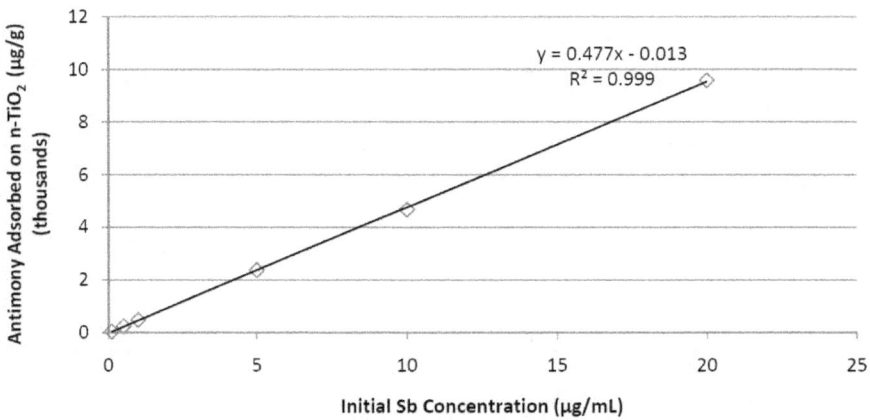

Fig. 3.5a: n-TiO$_2$ antimony adsorption capacity as a result of initial antimony concentration

Fig. 3.5b: Equilibrium antimony concentration dependence on initial antimony concentration

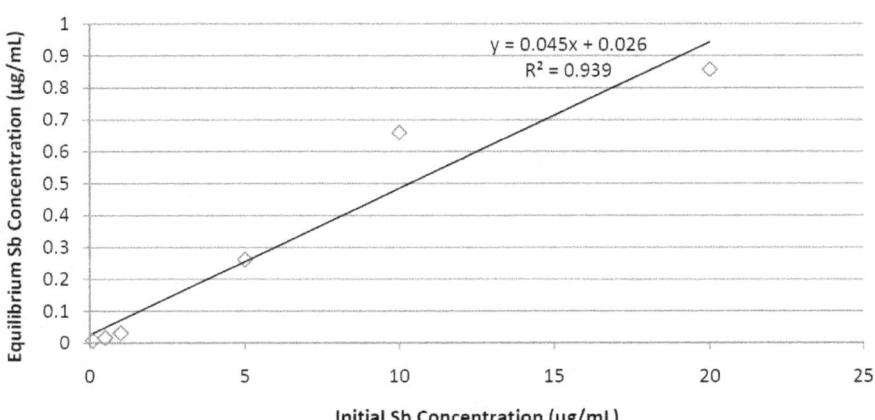

3.6.0 Investigating Sb(V) adsorption by micro- and nanoparticle titanium dioxide; effects on speciation; effects of light

A series of experiments was set up to determine the effect of nanoparticle and microparticle titanium dioxide on antimony speciation and adsorption. Furthermore, since nanoparticle titanium dioxide is known to be UV active and antimony (III) is easily oxidized in the presence of light, an experiment was set up comparing the effects of light or darkness on antimony speciation and adsorption by nanoparticle and microparticle titanium dioxide. All experiments used a 2 mg/mL concentration of adsorbent, and an antimony concentration of 50 ng/mL. All experiments were run in replicate to assure consistent results. An equilibration period of 48 hours was permitted before analysis by coupled HPLC – ICP – MS, using an approach published by Steely, et al (2007).

Since Sb(V) is already in a highly-oxidized state, the effects of light on adsorption were deemed to be negligible. In experiments utilizing Sb(V), the effect of light exposure on speciation was not investigated. Sb(III), however, is light sensitive; light has been found to promote the oxidation of Sb(III) to Sb(V). As a result, experiments involving Sb(III) were performed in both light and dark conditions. Even in dark conditions, however, some oxidation of Sb(III) was inevitable (Figs. 3.6e – 3.6f and 3.6i – 3.6j).

In the case of exposure of 50 ng/mL Sb(V) to 2 mg/mL nanoparticle titanium dioxide, after a 48 hour equilibration period, no detectable amount of antimony remains in solution (see figures 3.6a – 3.6b). This indicates complete adsorption of Sb(V) species by n-TiO$_2$. If microparticle titanium dioxide is used in place of nanoparticle titanium dioxide, some Sb(V) adsorption is evident (see Figs. 3.6c – 3.6d), but it is not complete (as in the case of the nanoparticle titanium dioxide); the concentration of antimony in the solution drops by approximately 50 percent (see Table 3.6). This corresponds to an antimony adsorption capacity of 25 µg Sb/g µ-TiO$_2$.

Sb(III) samples (50 ng/mL) that were kept in both dark and light, and exposed to 2 mg/mL nanoparticle titanium dioxide, exhibited complete adsorption to nanoparticle titanium dioxide. (see figures 3.6.0e – 3.6.0h). As a result, any oxidation of Sb(III) to Sb(V) is not noticeable, since all analyte is adsorbed. Sb(III) oxidation could be easily verified by raising the Sb(III) concentration, or by lowering the concentration of nanoparticle titanium dioxide. Sb(III) samples (50 ng/mL) kept in the dark and exposed to 2 mg/mL microparticle titanium dioxide exhibited oxidation to Sb(V), along with a degree of adsorption (see Figs. 3.6i – 3.6j). Sb(III) samples exposed to light behaved similarly to those kept in darkness, and were also oxidized to Sb(V) (figures 3.6k – 3.6l). Interestingly, however, those samples exposed to light demonstrate a greater total antimony adsorption. In the identical experiment performed using nanoparticle titanium dioxide, the Sb(III) was completely adsorbed regardless of the presence or absence of light.

In all cases, it appears that nanoparticle titanium dioxide is a better adsorbent of antimony than microparticle titanium dioxide. This is expected, as in general, nanoparticle adsorbents function better than their microparticle counterparts (often as a result of the much higher surface area). Intriguingly, Sb(III) exposed to microparticle titanium dioxide was oxidized to Sb(V), and was partially adsorbed. Since all antimony was adsorbed by the nanoparticle titanium dioxide, it is suspected that it was oxidized from Sb(III) to Sb(V) before adsorption, but this could not be definitively proven based on the experimental results.

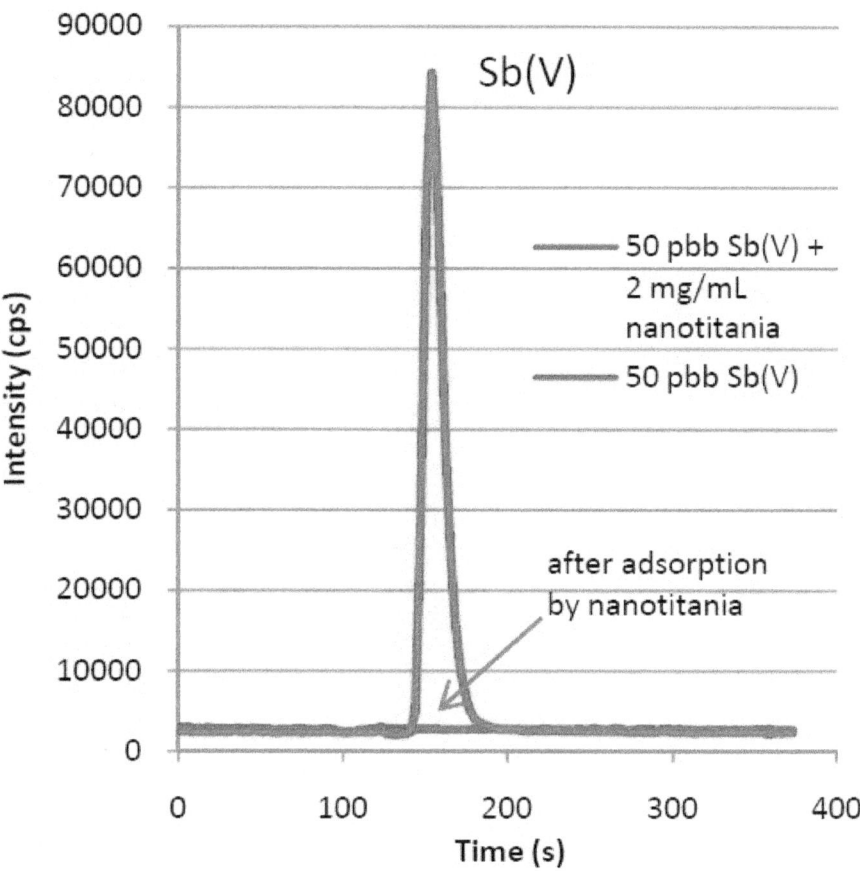

Fig. 3.6a: Plot of the Sb(V) (conc. = 50 ng/mL) adsorption capability of nanoparticle titanium dioxide (2 mg/mL) in light. Trial 1.

Fig. 3.6b: Plot of the Sb(V) (conc. = 50 ng/mL) adsorption capability of nanoparticle titanium dioxide (2 mg/mL) in light. Trial 2.

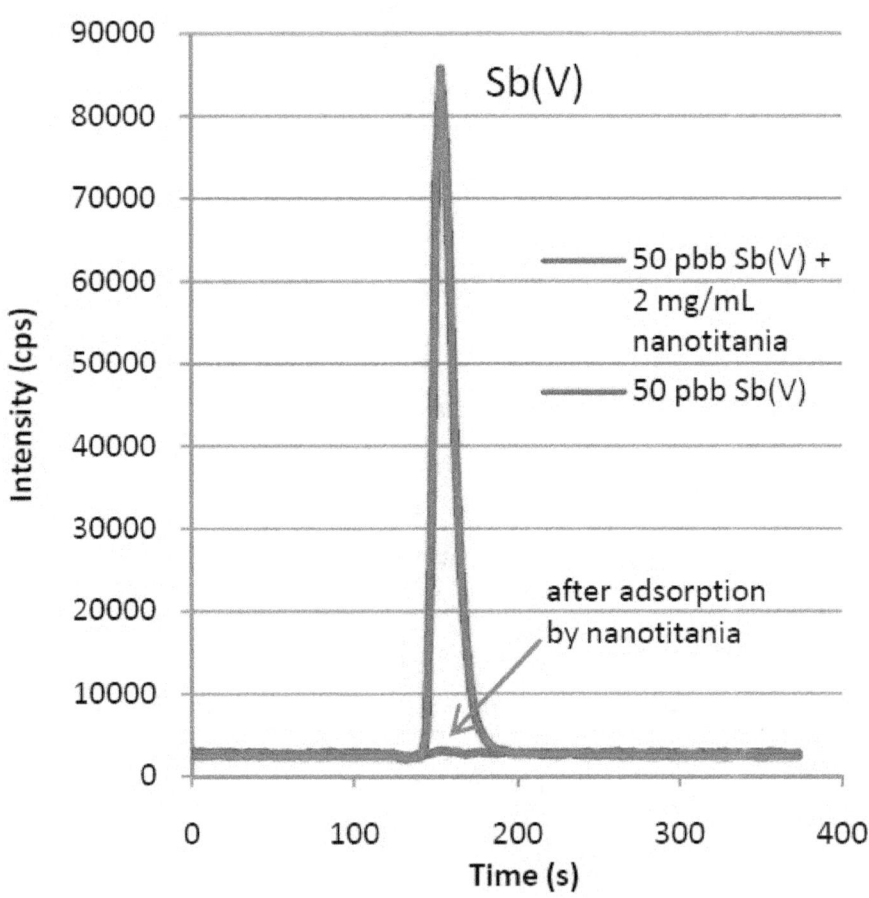

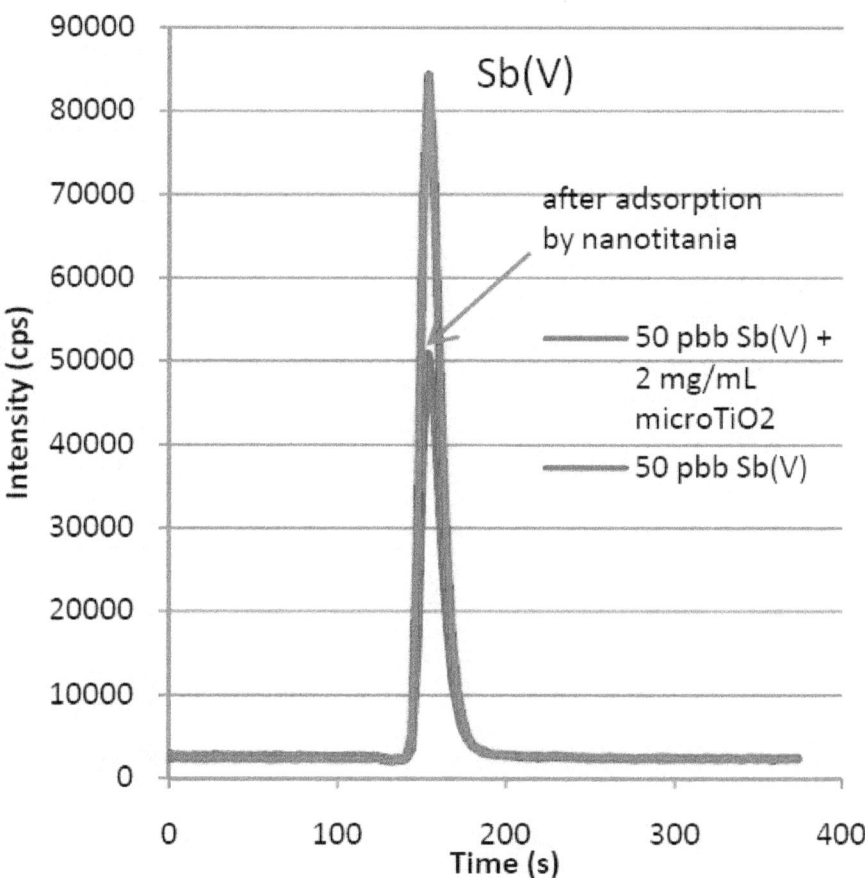

Fig. 3.6c: Plot of the Sb(V) (conc. = 50 ng/mL) adsorption capability of microparticle titanium dioxide (2 mg/mL) in light. Trial 1.

Fig. 3.6d: Plot of the Sb(V) (conc. = 50 ng/mL) adsorption capability of microparticle titanium dioxide (2 mg/mL) in light. Trial 2.

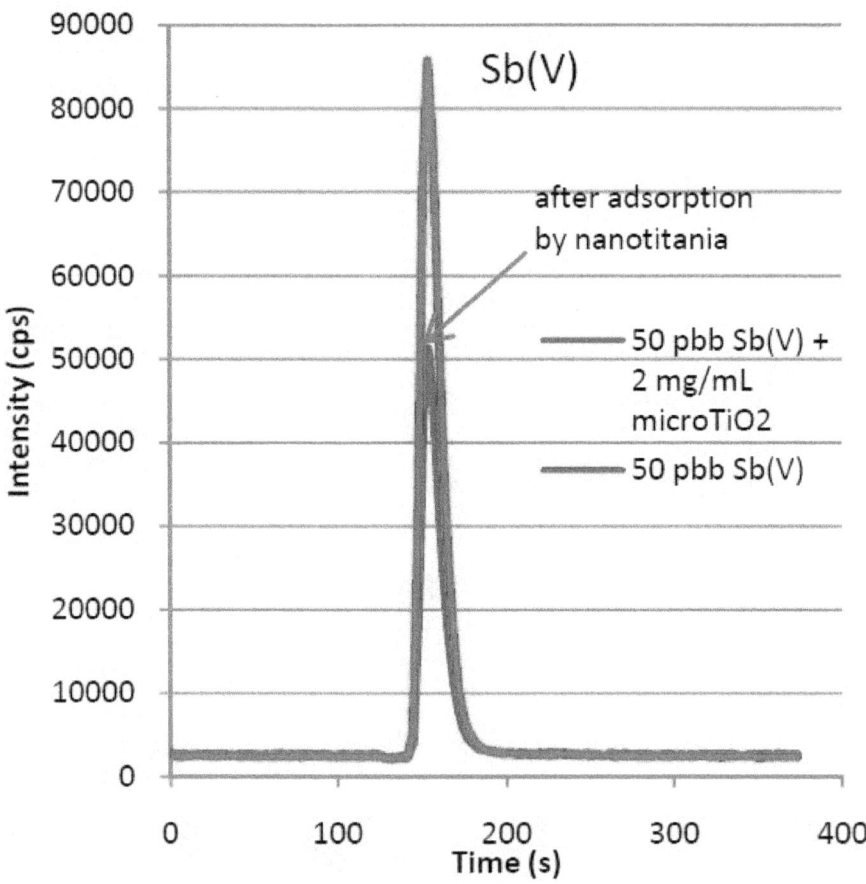

Fig. 3.6e: Plot of the Sb(III) (conc. = 50 ng/mL) adsorption capability of nanoparticle titanium dioxide (2 mg/mL) in darkness. Trial 1.

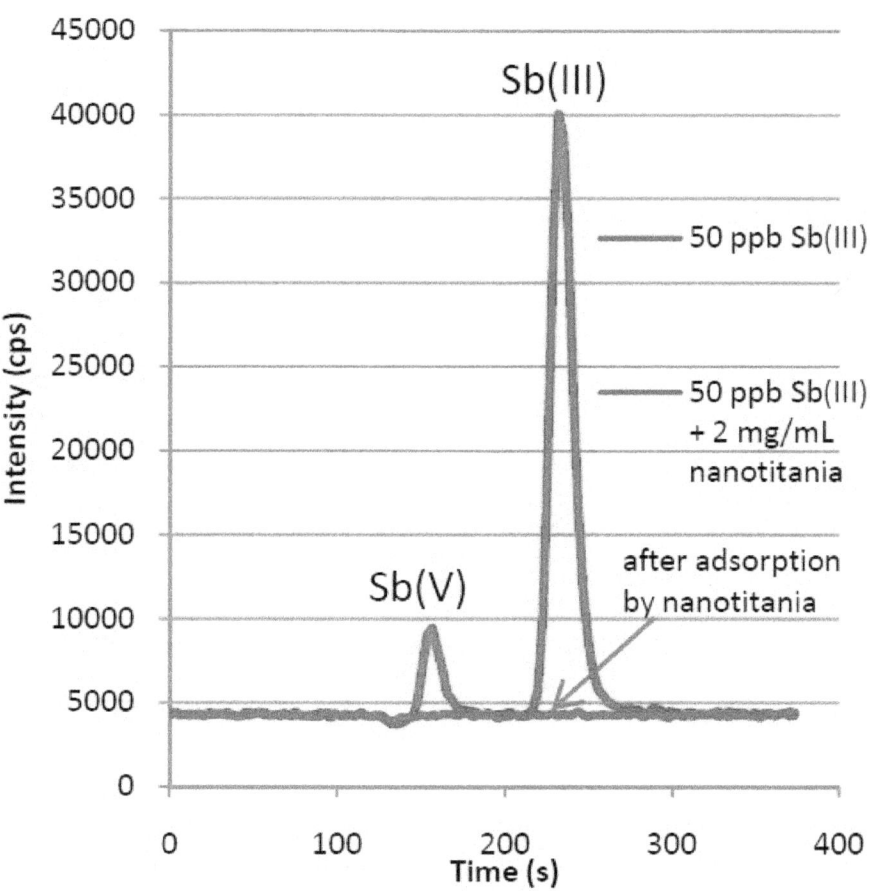

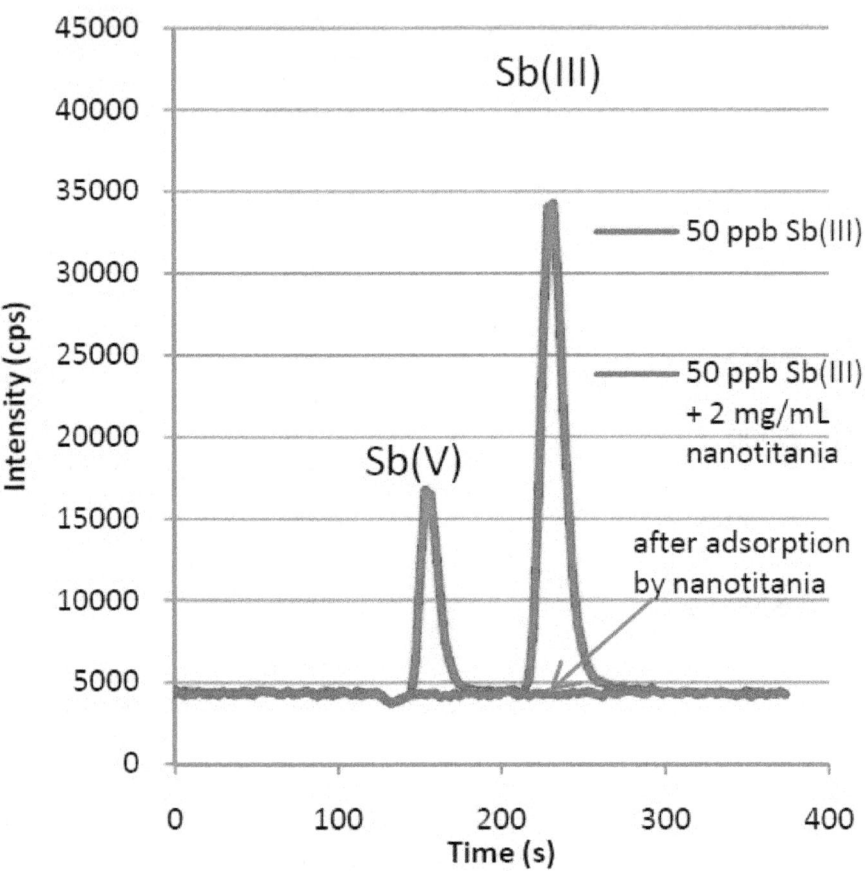

Fig. 3.6f: Plot of the Sb(III) (conc. = 50 ng/mL) adsorption capability of nanoparticle titanium dioxide (2 mg/mL) in darkness. Trial 2.

Fig. 3.6g: Plot of the Sb(III) (conc. = 50 ng/mL) adsorption capability of nanoparticle titanium dioxide (2 mg/mL) in light. Trial 1.

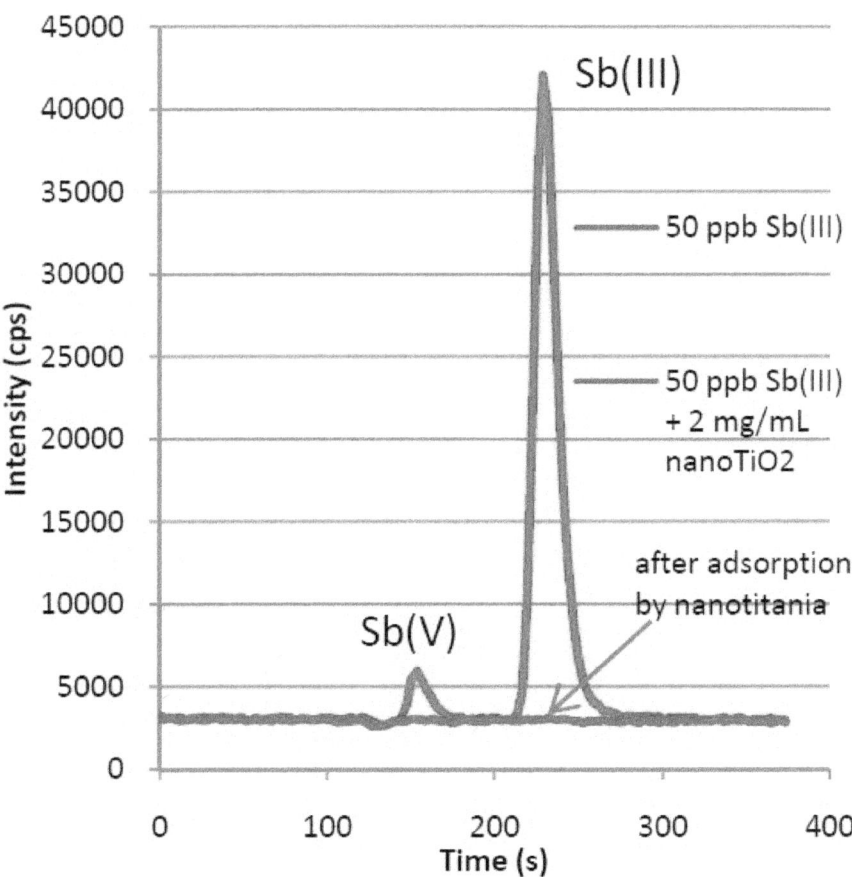

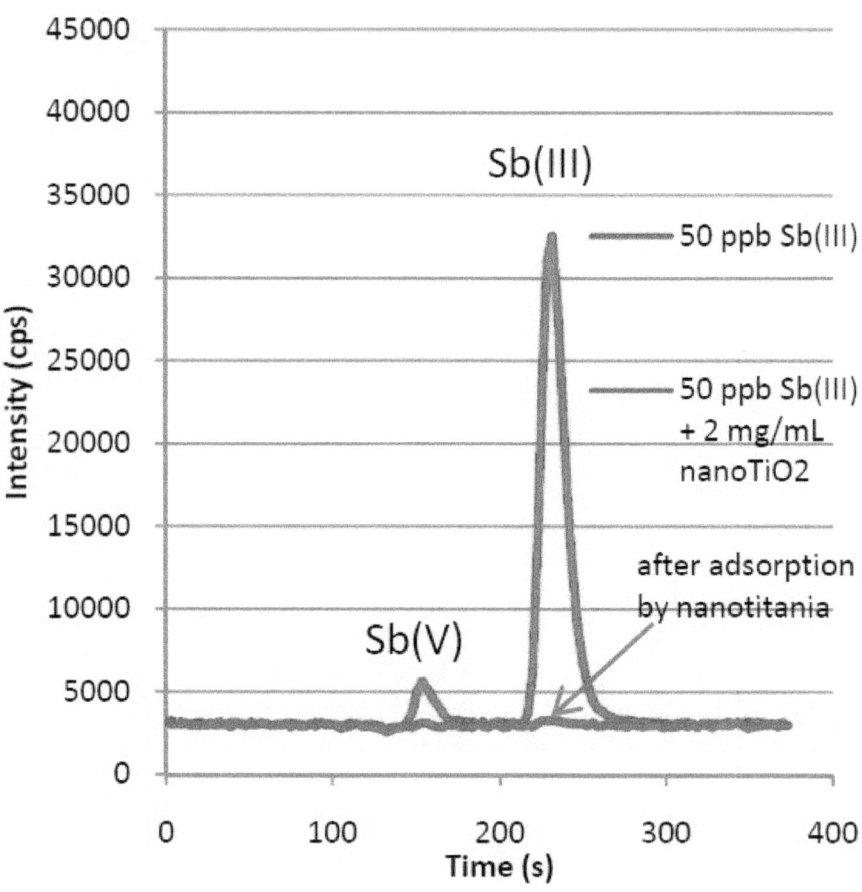

Fig. 3.6h: Plot of the Sb(III) (conc. = 50 ng/mL) adsorption capability of nanoparticle titanium dioxide (2 mg/mL) in light. Trial 2.

Fig. 3.6i: Plot of the Sb(III) (conc. = 50 ng/mL) adsorption capability of microparticle titanium dioxide (2 mg/mL) in darkness. Trial 1.

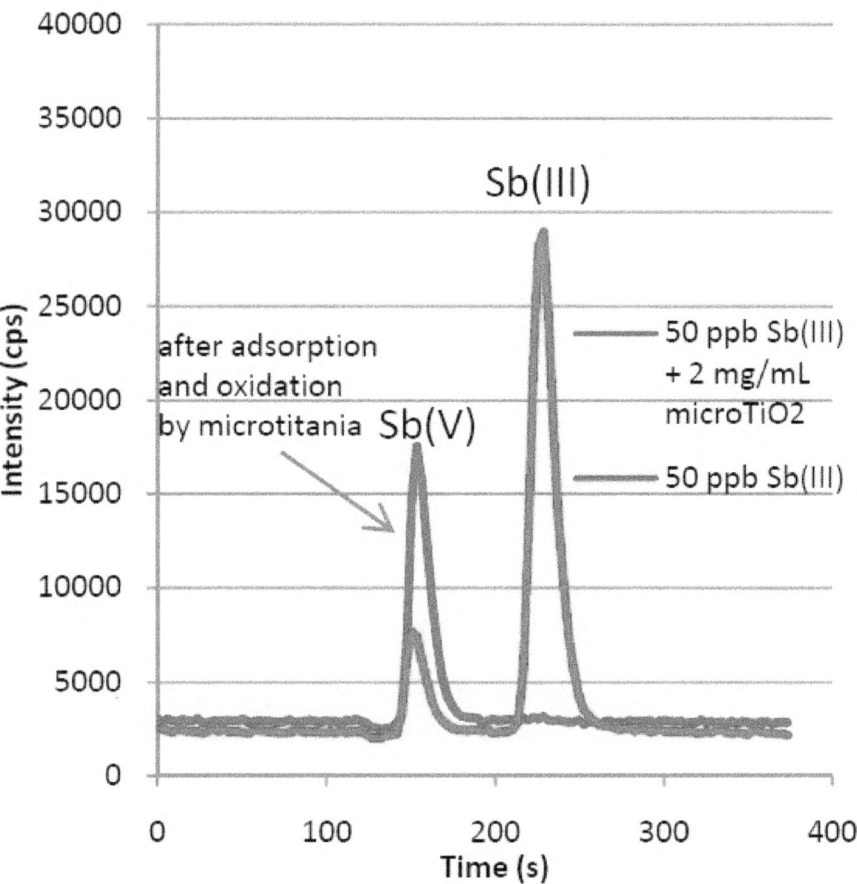

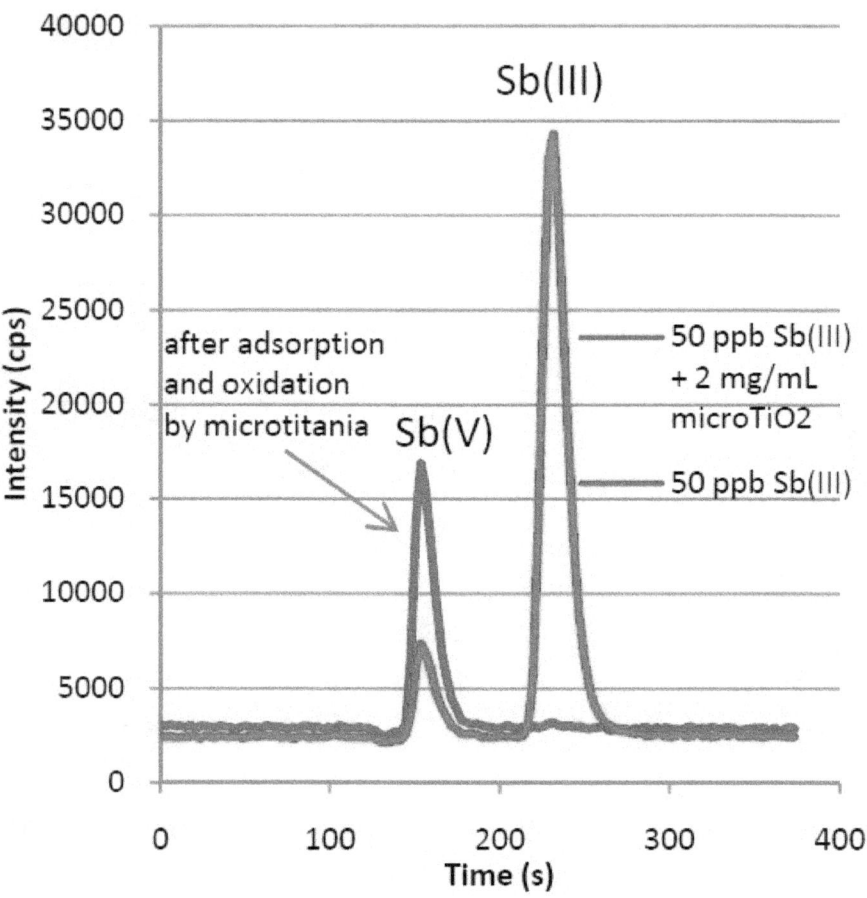

Fig. 3.6j: Plot of the Sb(III) (conc. = 50 ng/mL) adsorption capability of microparticle titanium dioxide (2 mg/mL) in darkness. Trial 2.

Fig. 3.6k: Plot of the Sb(III) (conc. = 50 ng/mL) adsorption capability of microparticle titanium dioxide (2 mg/mL) in light. Trial 1.

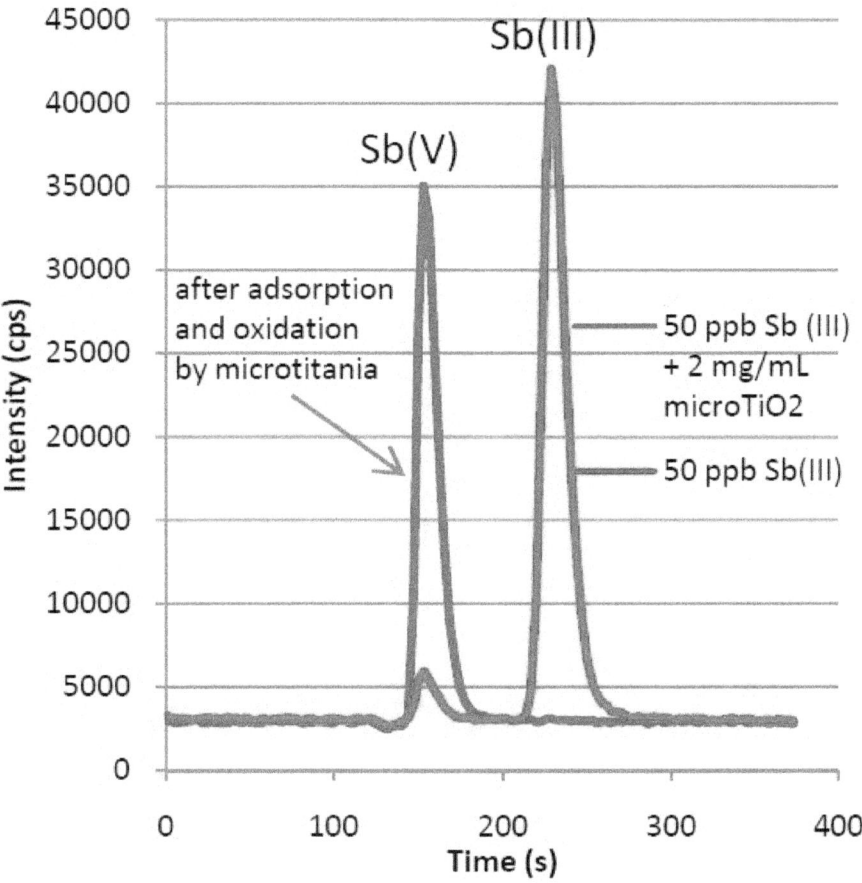

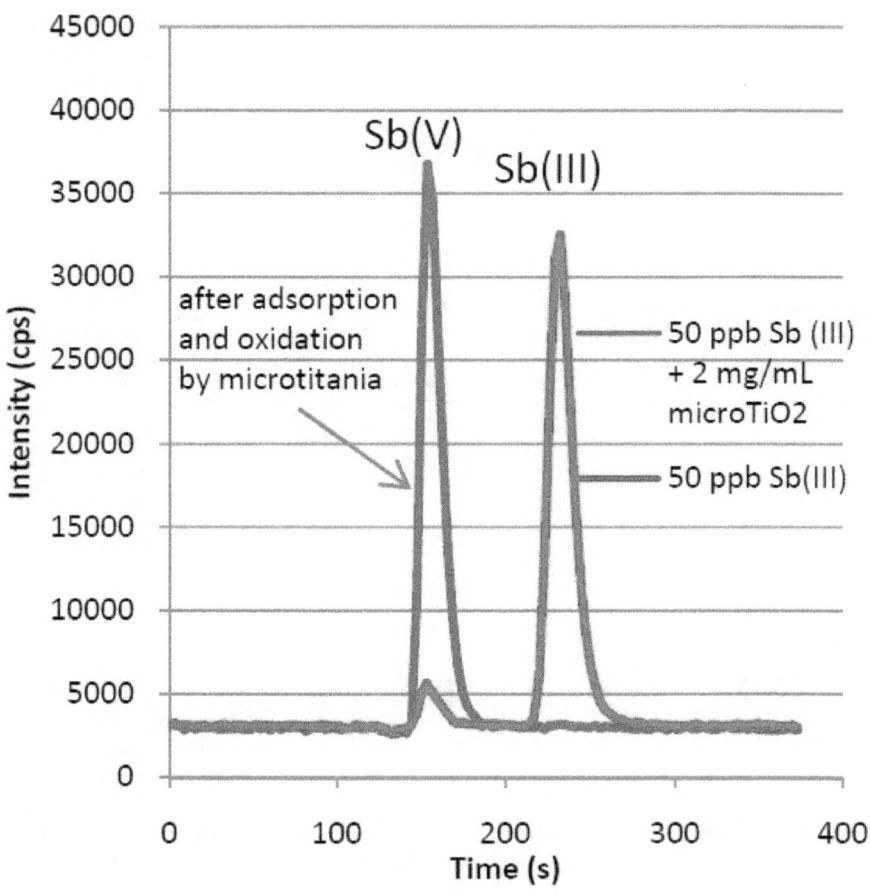

Fig. 3.6l: Plot of the Sb(III) (conc. = 50 ng/mL) adsorption capability of microparticle titanium dioxide (2 mg/mL) in light. Trial 2.

Table 3.6: Summary of Sb(III) and Sb(V) adsorption by micro- and nanoparticle titanium dioxide.

Sample	Initial Sb(III) Conc. (ng/mL)	Initial Sb(V) conc. (ng/mL)	Final Sb(III) conc. (ng/mL)	Final Sb(V) conc. (ng/mL)	Sb(III) – Sb(V) Percent Conversion (%)
50 ng/mL Sb(III) + 2 mg/mL μ-TiO$_2$, darkness, Trial 1	50	0	0	2.5	5.0
50 ng/mL Sb(III) + 2 mg/mL μ-TiO$_2$, darkness, Trial 2	50	0	0	2.4	4.8
50 ng/mL Sb(III) + 2 mg/mL μ-TiO$_2$, light, Trial 1	50	0	0	17.0	34.0
50 ng/mL Sb(III) + 2 mg/mL μ-TiO$_2$, light, Trial 2	50	0	0	18.1	36.2
50 ng/mL Sb(III) + 2 mg/mL n-TiO$_2$, darkness, Trial 1	50	0	0	0	0
50 ng/mL Sb(III) + 2 mg/mL n-TiO$_2$, darkness, Trial 2	50	0	0	0	0

50 ng/mL Sb(III) + 2 mg/mL n-TiO$_2$, light, Trial 1	50	0	0	0	0
50 ng/mL Sb(III) + 2 mg/mL n-TiO$_2$, light, Trial 2	50	0	0	0	0
50 ng/mL Sb(V) + 2 mg/mL µ-TiO$_2$, light, Trial 1	0	50	0	25.1	50.2
50 ng/mL Sb(V) + 2 mg/mL µ-TiO$_2$, light, Trial 2	0	50	0	25.2	50.4
50 ng/mL Sb(V) + 2 mg/mL n-TiO$_2$, light, Trial 1	0	50	0	0	0
50 ng/mL Sb(V) + 2 mg/mL n-TiO$_2$, light, Trial 2	0	50	0	0	0

3.7.0 Infrared spectra of pristine n-TiO$_2$ and n-TiO$_2$ containing adsorbed antimony

Following the precedent of Peta et al., (2006), infrared spectroscopy was used to collect spectra of both pristine nanoparticle titanium dioxide, and of nanoparticle titanium dioxide containing adsorbed antimony (~10 µg Sb/mg n-TiO$_2$). Resultant spectra display noticeable differences between the spectrum of the pristine nanoparticle titanium dioxide, and that containing adsorbed antimony (see figures 3.7a – 3.7b). Additionally, Peta et al. reported IR stretches at 830 cm^{-1} and 808 cm^{-1} for non-complexed As-O stretches and complexed As-O-Ti stretches, respectively. While the IR bands can't be expected to be identical for similar complexes formed by antimony, there are absorbance peaks evident in the general region (see figures 3.7d – 3.7d) present in the n-TiO$_2$/Sb sample that are not present in the pristine nanoparticle titanium dioxide. The absorbance region for the antimony oxides is roughly 700 – 800 cm^{-1} (Pitzer, 1990), and thus figures 3.7d – 3.7d suggest that the antimony is bound to the oxygen atoms of the nanoparticle titanium dioxide.

Figure 3.7a: DRIFT spectra of pristine n-TiO$_2$ and n-TiO$_2$ with adsorbed antimony. Trial 1.

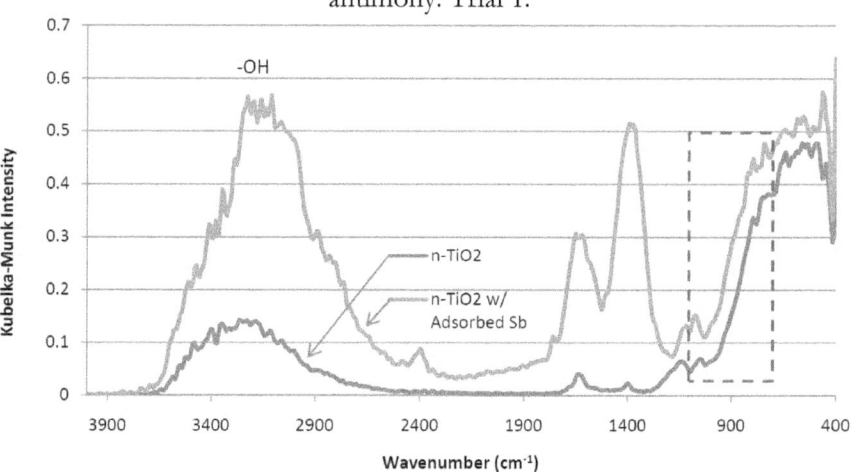

Figure 3.7b: DRIFT spectra of pristine n-TiO$_2$ and n-TiO$_2$ with adsorbed antimony. Trial 2.

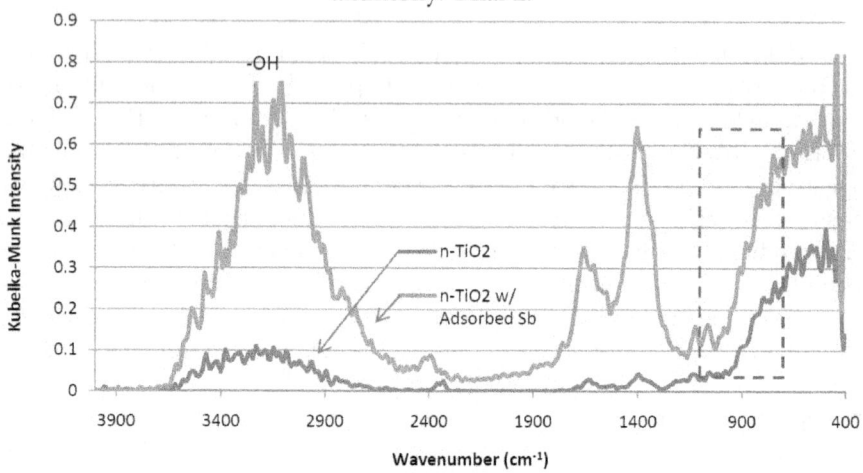

Figure 3.7c: Close-up DRIFT spectra of pristine n-TiO$_2$ and n-TiO$_2$ with adsorbed antimony. Trial 1.

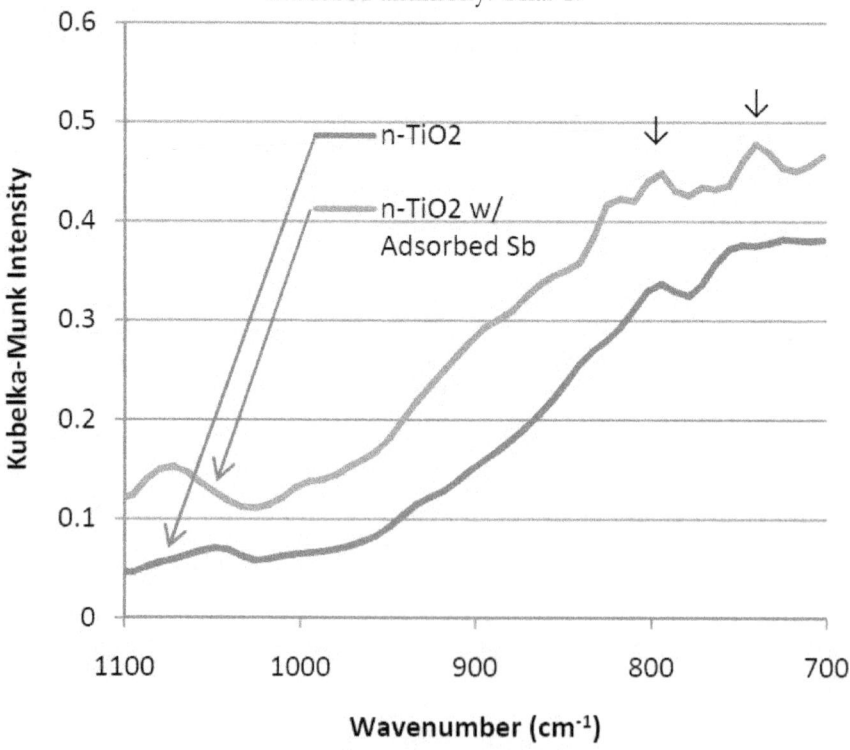

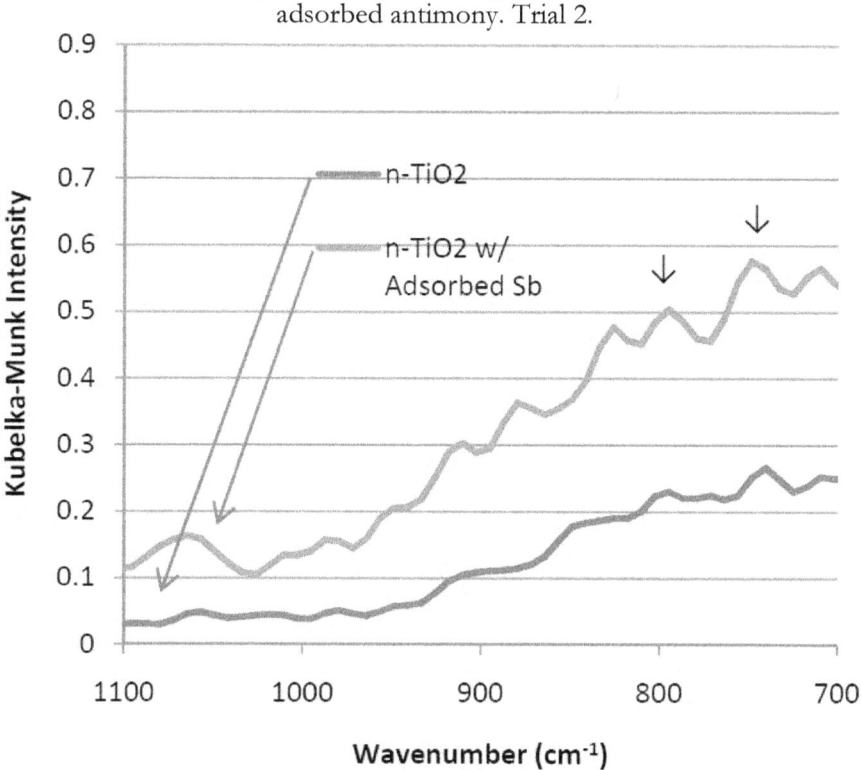

Figure 3.7d: Close-up DRIFT spectra of pristine n-TiO$_2$ and n-TiO$_2$ with adsorbed antimony. Trial 2.

3.8.0 Summary

At low pH (~1.4), there appears to be little difference between the antimony, cadmium, and lead adsorption capability of pristine and functionalized multi-walled carbon nanotubes. When comparing the antimony, cadmium, and lead adsorption capability of a wider range of materials (pristine MWCNTs, activated carbon, μ-TiO$_2$ and n-TiO$_2$), it was found that the best adsorbent of all three trace elements was n-TiO$_2$, followed by AC, μ-TiO$_2$, and MWCNTs. Nanoparticle titanium dioxide is a better adsorbent of antimony (both (III) and (V)) when compared to microparticle titanium dioxide. The effect of ambient light had little effect on the adsorption of antimony. DRIFT spectra of pristine nanoparticle titanium dioxide, and titanium dioxide containing adsorbed antimony reveal that the Sb-O absorbance peak may be evident at approximately 794 cm^{-1}.

4.0 CONCLUSION

Given the increasing importance of nanotechnology, it can be expected that products containing nanomaterials will become ever more ubiquitous in the surrounding environment. Since it is unlikely that the commercialization of useful nanotechnological devices will cease, it is of utmost importance that interactions between nanomaterials and the environment be studied preemptively.

The size of nanomaterials allows many of them to pass uninhibited into the interior of cells. This, when coupled with the facile adsorption of toxins, suggests the importance of studying the adsorption of toxic materials by nanomaterials. Cadmium, lead, and antimony are all toxic to mammalian cells, and when adsorbed to a nanomaterial, the transportability, and thus toxicity of the material is increased.

For two common nanomaterials, multi-walled carbon nanotubes (MWCNTs) and nanoparticle titanium dioxide (n-TiO_2), a series of comparative adsorption experiments was performed. Additionally, the adsorption capacity of the MWCNTs was compared to that of oxidized MWCNTs, and activated carbon (AC) (a well-characterized adsorbent, as a control). The adsorption capacity of nanoparticle titanium dioxide, while being compared to the carbon-based adsorbents, was also compared to microparticle titanium dioxide (μ-TiO_2). Additionally, the effect of diffuse light and adsorption on antimony speciation was studied.

Results indicate that, of the adsorbents studied, nanoparticle titanium dioxide is the best adsorbent of cadmium, lead, and antimony, followed by activated carbon, microparticle titanium dioxide, surface oxidized MWCNTs, and pristine MWCNTs. At pH ~1.4, there was no great difference between the cadmium, lead, or antimony adsorption capacity of functionalized and pristine MWCNTs. The adsorption capacity of pristine MWCNTs, however, did increase slightly when the solution pH was raised from ~1.4 to ~6.4 (from ~0% adsorption to ~10% adsorption of available antimony in an initial 50 ppm Sb solution). For all adsorbents, at all pHs studied, both the Langmuir and Freundlich equations provided good fits of the data. In general, however, the fits for MWCNTs and microparticle titanium dioxide were not as satisfactory as those for AC and nanoparticle titanium dioxide. The calculated Langmuir parameters (Table 3.4b) suggest that the cadmium monolayer adsorption capacity is higher than that of antimony or lead, for all adsorbents except activated carbon. This observation has some limitations, as the Langmuir model does not account for the possibility of multilayer adsorption. Additionally, the Langmuir parameters suggest that cadmium and lead are most strongly adsorbed by activated carbon, but that antimony is most strongly adsorbed by microparticle titanium dioxide. Freundlich parameters imply that cadmium and lead are most strongly adsorbed by nanoparticle titanium dioxide, but that antimony is bound most strongly (energetically) to the pristine MWCNTs. Once again, the interpretation of these results is limited, given the simplistic assumptions made by Freundlich adsorption theory.

Results of the antimony speciation studies indicate that microparticle titanium dioxide (μ-TiO_2) adsorbed the Sb(III) in solution, which was then oxidized to Sb(V). In the case of nanoparticle titanium dioxide (n-TiO_2), it is unknown if Sb(III) was oxidized to Sb(V) in solution, as all antimony was completely adsorbed by the nanoparticles at the concentrations studied. This may be overcome in future experiments by raising the concentration of antimony, or lowering the concentration of the nanoparticle titanium dioxide. The presence of ambient light had only a small effect on adsorption and oxidation; antimony adsorption by microparticle titanium dioxide was more complete in the absence of light.

Finally, diffuse reflectance infrared Fourier transform (DRIFT) spectroscopy of both pristine nanoparticle titanium dioxide, and titanium dioxide containing adsorbed antimony, revealed the presence of absorbance peaks in the 700 – 800 cm^{-1} region. This is the absorbance region for the antimony oxides (Pitzer, 1990), it is thus suggested that the antimony is bound to the oxygen atoms of the nanoparticle titanium dioxide.

Overall, it appears that nanoparticle titanium dioxide is a good adsorbent for lead, cadmium, and antimony. Pristine MWCNTs, when compared to the other adsorbents studied (activated carbon, μ-TiO_2 and n-TiO_2), were found to possess the poorest adsorption of the analytes utilized. As has been reported in other work, functionalized MWCNTs generally are better adsorbents of trace metal analytes when compared to pristine MWCNTs. Thus, while this research has found that pristine MWCNTs adsorb antimony, cadmium, and lead only poorly, the possibility of environmental oxidation of MWCNTs suggests that adsorbent-adsorbate interactions may be stronger in real-world situations. It is therefore difficult to adequately assess the environmental impact of MWCNTs.

Toxicity studies have indicated that nanoparticle titanium dioxide is not significantly toxic when in solution. Nanoparticle titanium dioxide was found to be a superior adsorbent of cadmium, lead, and antimony, but it is unknown if titanium dioxide nanoparticles containing adsorbed toxic metals would be, in practice, any more toxic than pristine titanium dioxide nanoparticles. The cellular transport of titanium dioxide nanoparticles containing adsorbed toxic elements is a possibility. The strength with which cadmium, lead, and antimony are bound to the nanoparticles is unknown, so it is difficult to determine if the presence of nanoparticle titanium dioxide would prove detrimental to a localized environment.

4.1.0 Future Work

There remains a large amount of research to be completed regarding trace element adsorption by multi-walled carbon nanotubes and nanoparticle titanium dioxide. More experiments need to be done exploring the adsorption of antimony by surface functionalized MWCNTs. A consistent protocol for the surface oxidation of the carbon nanotubes needs to be created, and verified as working. The surface of oxidized carbon nanotubes should be characterized using a technique such as x-ray photoelectron spectroscopy (XPS) or Boehm's titration. Using surface oxidized MWCNTs, a plot of antimony adsorption versus solution pH needs to be created. Additionally, the effect of pH on antimony adsorption by nanoparticle titanium dioxide and pristine MWCNTs also needs to be studied more in depth.

In general, the analyte concentrations studied should be increased. In the case of lead adsorption by nanoparticle titanium dioxide, for instance, the adsorption isotherms indicated that the lead was effectively completely adsorbed by the Nanoparticle titanium dioxide, which resulted in an incomplete isotherm. In place of increasing the analyte concentrations, it may be easier to simply decrease the concentration of the adsorbent.

Finally, future adsorption data can be fit using additional isotherms (Toth, Brunauer-Emmett-Teller) not studied at this time. In the case of better isotherm fits, more accurate fitting parameters may be able to be extracted.

5.0 REFERENCES

Aitchison, T., Ginic-Markovic, M., Matisons, J., Simon, G., Fredericks, P. Purification, Cutting, and Sidewall Functionalization of Multi-walled Carbon Nanotubes Using Potassium Permanganate Solutions. Journal of Physical Chemistry C. Vol. 111, Issue 6. 2007. pp. 2440 – 2446.

Basiuka, V., Albarrana, G., Basiuka, E., Saniger, J. Stability of interstellar fullerenes under high- dose γ-irradiation: new data. Advances in Space Research. Vol. 36, Issue 2. 2005. pp. 173-177.

Baughman, R. H., Zakhidov, A. A. and De Heer, W. A. Carbon nanotubes – the route toward applications. Science, Vol. 297, Issue 5582. 2002. pp. 787-792.

Baughman, R., Cui, C., Zakhidov, A., Iqbal, Z., Barisci, J., Spinks, G., Wallace, G., Mazzoldi, A., De Rossi, D., Rinzler, A., Jaschinski, O., Roth, S., Kertesz, M. Carbon Nanotube Actuators. Science. Vol. 284, No. 5418. 1999. pp. 1340-1344.

Boehm, H., Diehl, E., Heck, W., Sappok, R. Surface oxides of carbon. Angewandte Chemie International Edition. Vol. 3. 1964. pp. 669.

Bolstera, C., and Hornbergerb, G. On the Use of Linearized Langmuir Equations. Soil Science Society of America Journal. Vol. 71. 2007. pp. 1796-1806.

Bottini, M., Bruckner, S., Nika, K., Bottini, N., Bellucci, S., Magrini, A., Bergamaschi, A., Mustelin, T. Multi-walled carbon nanotubes induce T lymphocyte apoptosis. Toxicology Letters. Vol. 160, Issue 2. 2006. pp. 121-126.

Brumfiel, G. Nanotechnology: A little knowledge.... Nature, Vol. 424. 2003. pp. 246-248.

Brunauer, S., Emmett, P., Teller, E. Adsorption of Gases in Multimolecular Layers. Journal of the American Chemical Society. Vol. 60, No. 2. 1938. pp. 309-319.

Cataldo, F., Compagnini, G., and D'Urso, L. Adsorption of dinitrogen tetroxide (N2O4) on multi-walled carbon nanotubes (MWCNTs). Fullerenes, Nanotubes, and Carbon Nanostructures, Vol. 16, Issue 2. 2008. pp. 154-164.

Castro, M., and Schmidt, H. K. Transparent conducting antimony-doped tin oxide films containing functionalized multi-walled carbon nanotubes. Physica Status Solidi A, Vol. 204, Issue 10. 2007. pp. 3380-3386.

Chang, H., Lee, J. D., Lee, S. M., Lee, Y. H. Adsorption of NH3 and NO2 molecules on carbon nanotubes. Applied Physics Letters. Vol. 79. Iss. 23. 2001. pp. 3863-3865.

Chen, C. L., Hu, J., and Xu, D. Surface complexation modeling of Sr(II) and Eu(III) adsorption onto oxidized multiwall carbon nanotubes. Journal of Colloid and Interface Science, Vol. 323, Issue 1. 2008. pp. 33-41.

Chen, L. H., and Yang, J. L. Acute toxicity of antimony chloride and its effects on oxygen consumption of common carp (Cyprinus carpio). Bulletin of Environmental Contamination and Toxicology, Vol. 78, Issue 6. 2007. pp. 459-462.

Chen, W., Duan, L., and Zhu, D. Q. Adsorption of polar and nonpolar organic chemicals to carbon nanotubes. Environmental Science and Technology, Vol. 41, Issue 24. 2007. pp. 8295-8300.

Chiou, C. Partition and Adsorption of Organic Contaminants in Environmental Systems. Wiley- IEEE, 2002. pp. 41-43.

Chowdhury, U. K., Bhajan, B. K., Chowdhury, T. R., Samanta, G., Mandal, B. K., Basu, G. C., Chanda, C. R., Lodh, D., Saha, K. C., Mukherjee, S.K., Roy, S., Kabir, S., Quamruzzaman, Q. and Chakraborti, D. Groundwater arsenic contamination in Bangladesh and West Bengal, India. Environmental Health Perspectives, Vol. 108, Issue 5. 2000. pp. 393-397.

Collins, P., Phaedon, A. "Nanotubes for Electronics". Scientific American. December 2000. pp. 63-69.

Colvin, V. The potential environmental impact of engineered nanomaterials. Nature Biotechnology. Vol. 21, No. 10. 2003. pp. 1166-1170.

Dreher, K. Health and Environmental Impact of Nanotechnology: Toxicological Assessment of Manufactured Nanoparticles. Toxicological Sciences. Vol. 77. 2004. pp. 3-5.

Dresselhaus, M., Dresselhaus, G., Avouris, P. Carbon Nanotubes: Synthesis, Structure, Properties and Applications. 2000. Springer-Verlag.

Duan, H., McKinnon, J. Nanoclusters produced in flames. Journal of Physical Chemistry. Vol. 96, Issue 49. 1994. pp. 12815-12818.

Ebbesen, T., Ajayan, P. Large-scale synthesis of carbon nanotubes. Nature. Vol. 358. 1992. pp. 220-222.

Federici, G., Shaw, B., Handy, R. Toxicity of titanium dioxide nanoparticles to rainbow trout (Oncorhynchus mykiss): Gill injury, oxidative stress, and other physiological effects. Aquatic Toxicology. Vol. 84, Issue 4. 2007. pp. 415-430.

Freundlich, H. Colloid and Capillary Chemistry. Methuen, London, 1926.

Giammar, D., Maus, C., Xie, L. Effects of Particle Size and Crystalline Phase on Lead Adsorption to Titanium Dioxide Nanoparticles. Environmental Engineering Science. Vol. 24. 2007. pp. 85-95.

Girgis, G., Scott, P., Schulert, A.Browne. Acute tolerance of mice to tartar emetic. Toxicology and Applied Pharmacology. Vol. 7, Issue 5. 1965. pp. 727-731.

Gebel, T., Christensen, S., Dunkelberg H. Comparative and environmental genotoxicity of antimony and arsenic. Anticancer Research. Vol. 17. 1997. pp. 2603-2607.

Gao, Y., Masuda, Y., Seo, W., Ohta, H., Koumoto, K. TiO2 nanoparticles prepared using an aqueous peroxotitanate solution. Ceramics International. Vol. 30, Issue 7. 2004. pp. 1365-1368.

Goncalves, J. E., Gushikem, Y., and de Castro, S. C. Preparation and properties of antimony (V) oxide adsorbed on silica-titanium dioxide mixed oxide. Journal of Non-crystalline Solids, Vol. 260, Issue 1. 1999. pp. 125-131.

Greenwood, N., Earnshaw, A. Chemistry of the Elements, 2nd Edition. 1997. Oxford: Butterworth-Heinemann.

Guo, J., Lundstrom, M., Datta, S. Performance projections for ballistic carbon nanotube field- effect transistors. Applied Physics Letters. Vol. 80. 2002. p. 3192.

Helland, A., Scheringer, M., Siegrist, M., Kastenholz, H. G., Wiek, A. and Scholz, R. W. Risk assessment of engineered nanomaterials: A survey of industrial approaches. Environmental Science and Technology, Vol. 42, Issue 2. 2008. pp. 640-646.

Iijima, S. Helical microtubules of graphitic carbon. Nature. Vol. 354. 1991. pp. 56-58.

Iwasaki, M., Hara, M., Ito, S. Facile synthesis of nanocrystalline anatase particles from titanyl sulfate. Journal of Materials Science Letters. Vol. 17. 1998. pp. 1769-1771.

Jang, H., Kim, S. J., Kim, S. K. Effect of Particle Size and Phase Composition of Titanium Dioxide Nanoparticles on the Photocatalytic Properties. Journal of Nanoparticle Research. Vol. 3. No. 3. 2001. pp. 141-147

Jang, H., Kim, S. Controlled synthesis of titanium dioxide nanoparticles in a modified diffusion flame reactor. Materials Research Bulletin. Vol. 36, Issue 3. 2001. pp. 627-637.

Jiaa, X., Hea, W., Zhanga, X., N2 adsorption curve of mesoporous SiO_2: A neglected characterization technique. Materials Letters. Vol. 61, Issue 22. 2007. pp. 4377 – 4380.

Kajitvichyanukul, P., Chenthamarakshan, C. R., and Rajeshwar, K. Adsorption of thallium(I) ions on titanium dioxide particle surfaces in aqueous media. Adsorption Science and Technology, Vol. 21, Issue 3. 2003. pp. 217-228.

Kocabas, S., Kopac, T., and Dogu, G. Effect of thermal treatments and palladium loading on hydrogen sorption characteristics of single-walled carbon nanotubes. International Journal of Hydrogen Energy, Vol. 33, Issue 6. 2008. pp. 1693-1699.

Kolenko, Y., Burukhin, A., Churagulov, B., Oleynikov. Synthesis of nanocrystalline TiO2 powders from aqueous TiOSO4 solutions under hydrothermal conditions. Materials Letters. Vol. 57, Issue 5. 2003. pp. 1124-1129.

Kong, J., Cassell, A., Dai, H. Chemical vapor deposition of methane for single-walled carbon nanotubes. Chemical Physics Letters. Vol. 292. 1998. pp. 567-574.

Lam, C., James, J., McCluskey, R., Hunter, R. Pulmonary Toxicity of Single-Wall Carbon Nanotubes in Mice 7 and 90 Days After Intratracheal Instillation. Toxicological Sciences. Vol. 77. 2004. pp. 126-134.

Lee, K., Trochimowics, H., Reinhardt, C. Pulmonary response of rats exposed to titanium dioxide (TiO2) by inhalation for two years. Toxicology and Applied Pharmacology. Vol. 79, Issue 2. 1985. pp. 179-192.

Li, W., Xie, S., Qian, L., Chang, B., Zou, B., Zhou, W., Zhao, R., Wang, G. Large-Scale Synthesis of Aligned Carbon Nanotubes. Science. Vol. 274, No. 5293. 1996. pp. 1701-1703.

Li, Y., Hsieh, C. Synthesis of carbon nanotubes by combustion of a paraffin wax candle. Micro and Nano Letters. Vol. 2, Issue 3. 2007. pp. 63-66.

Li, Y., Wang, S., Luan, Z., Ding, J., Xu, Cailu., Wu, Dehai. Adsorption of cadmium(II) from aqueous solution by surface oxidized carbon nanotubes. Carbon. Vol. 41, Issue 5. 2003. pp. 1057-1062.

Li, Y., Wang, S., Wei, J., Zhang, X., Xu, C., Luan, Z., Wu, D., Wei, B. Lead Adsorption on Carbon Nanotubes. Chemical Physics Letters. Vol. 357. 2002. pp. 263-366.

Liu, J. Y., Guo, Z., and Meng, F. L. A novel antimony-carbon nanotube-tin oxide thin film: Carbon nanotubes as growth guider and energy buffer. Application for indoor air pollutants gas sensor. Journal of Physical Chemistry C, Vol. 112, Issue 15. 2008. pp. 6119-6125.

Linkov, I., Satterstrom, F. K., Stevens, J., Ferguson, E. and Pleus R. C. Multi-criteria decision analysis and environmental risk assessment for nanomaterials. Journal of Nanoparticle Research, Vol. 9, Issue 4. 2007. pp. 543-554.

Matott, L. IsoFit: Isotherm Fitting Tool. Groundwater Research Group. State University of New York, Buffalo. 2007. http://www.groundwater.buffalo.edu/software/IsoFit/IsoFitMain.html

MIT Institute for Soldier Nanotechnologies. Accessed Sept. 09, 2008. http://web.mit.edu/isn/

Monteiro-Riviere, N., Nemanich, R., Inman, A., Wang, Y., Riviere, J. Multi-walled carbon nanotube interactions with human epidermal keratinocytes. Toxicology Letters. Vol. 155, Issue 3. 2005. pp. 377-384.

Monthioux, M., Kuznetsov, V. Who should be given credit for the discovery of carbon nanotubes? Carbon. Vol. 44. 2006. pp. 1621-1622.

Mukhopadhyay, B., Lahiri, S. Adsorption de 125Sb on alumina and titanium dioxide surfaces. Journal of Radioanalytical and Nuclear Chemistry. Vol. 273, No. 2. 2007. pp. 423-426.

Murr, L. Nanoparticulate Materials in Antiquity: The Good, the Bad and the Ugly. Microscopy and Microanalysis. Vol. 13. 2007. pp. 1118-1119.

National Science Foundation. Societal Implications of Nanoscience and Nanotechnology. NSF Publications, March 2001.

Newton, K., Amarasiriwardena, D., Xing, B. Distribution of soil arsenic species, lead and arsenic bound to humic acid molar mass fractions in a contaminated apple orchard. Environmental Pollution. Vol. 143. 2006. pp. 197-205.

Newton, P. E., Bolte, H. F., and Daly, I. W. Subchronic and chronic inhalation toxicity of antimony trioxide in the rat. Fundamental and Applied Toxicology. Vol. 22, Issue 4. 1994. pp. 561-576.

Oberlin, A., Endo, M., Koyama, T. Filamentous growth of carbon through benzene decomposition. Journal of Crystal Growth. Vol. 32, Issue 3. 1976. pp. 335-349.

Oorts, K., Smolders, E., and Degryse, F. Solubility and toxicity of antimony trioxide (Sb2O3) in soil. Environmental Science and Technology, Vol. 42, Issue 12. 2008. pp. 4378-4383.

Pena, M., Meng, X., Korfiatis, G., Jing, C. Adsorption Mechanism of Arsenic on Nanocrystalline Titanium Dioxide. Environ. Sci. Technol. Vol. 40. 2006. pp. 1257-1262.

Peng, X., Luan, Z., Ding, J., Di, Z., Li, Y. and Tian, B. Ceria nanoparticles supported on carbon nanotubes for the removal of arsenate from water. Materials Letters, Vol. 59, Issue 4. 2003. pp. 399-403.

Piao, L. Y., Liu, Q. R., and Li, Y. D. Adsorption of L-phenylalanine on single-walled carbon nanotubes. Journal of Physical Chemistry C, Vol. 112, Issue 8. 2008. pp. 2857-2863.

Pitzer, E. FT-IR Analyses of the Thermal and Oxidative Decomposition Products of Antimony Sulfides. Applied Spectroscopy. Vol. 44, Issue 9. 1990. pp. 1498-1499.

Popov, V. Carbon nanotubes: properties and applications. Materials Science and Engineering: Reports. Vol. 43, Issue 3. 2004. pp. 61-102.

Reddya, K. M., Gopal, C. V., Narayanana, S., Rajab, P. R., and Chatterjib, P. R. Photostabilization of dye on anatase titanium dioxide nanoparticles by polymer capping. Journal of Physics and Chemistry of Solids, Vol. 63, Issue 1. 2002. pp. 135-143.

Ruiz-Haas, P., Amarasiriwardena, D., Xing, B. Determination of trace metals bound to soil humic acid species by size exclusion chromatography and inductively coupled plasma mass spectrometry. Humic Substances: Structures, Properties and Uses. Royal Society of Chemistry. 1998. pp. 147-163.

Sayes, C., Wahi, R., Kurian, P., Liu, Y., West, J., Ausman, K., Warheit, D., Colvin, V. Correlating Nanoscale Titanium dioxide Structure with Toxicity: A Cytotoxicity and Inflammatory Response Study with Human Dermal Fibroblasts and Human Lung Epithelial Cells. Toxicological Sciences. Vol. 92, Issue 1. 2006. pp. 174-185.

Schadler, L. S., Giannaris, S. C., and Ajayan, P. M. Load transfer in carbon nanotube epoxy composites. Applied Physics Letters, Vol. 73. 1998. pp. 3842-3844.

Silver, S., Budd, K., Leahy, K., Shaw, W., Hammond, D., Novick, R., Willsky, G., Malamy, M., Rosenberg, H. Inducible plasmid-determined resistance to arsenate, arsenite, and antimony (III) in escherichia coli and Staphylococcus aureus. Journal of Bacteriology. Vol. 146, No. 3. 1981. pp. 983-996.

Stafiej, A., and Pyrzynska, K. Adsorption of heavy metal ions with carbon nanotubes. Separation and Purification Technology, Vol. 58, Issue 1. 2007. pp. 49-52.

Steely, S., Amarasiriwardena, D., Xing, B. An investigation of inorganic antimony species and antimony associated with soil humic acid molar mass fractions in contaminated soils. Environmental Pollution. Vol. 148. 2007. pp. 590-598.

Stern, S. T. and McNeil, S. E. Nanotechnology safety concerns revisited. Toxicological Sciences, Vol. 101. 2008. pp. 4-21.

Tang, Z., Zhan, L., Wang, N., Zhang, X., Wen, G., Li, G., Wang, J., Chan, C., Sheng, P. Superconductivity in 4 Angstrom Single-Walled Carbon Nanotubes. Science. Vol. 292, No. 5526. 2001. pp. 2462-2465.

Tang, W., Santare, M., Advani, S. Melt processing and mechanical property characterization of multi-walled carbon nanotube/high density polyethylene (MWNT/HDPE) composite films. Carbon. Vol. 41, Issue 14. 2003. pp. 2779-2785.

Thomas, R., and Anirudhan, T. S. Equilibrium studies for the sorption of Hg(II) on silica-titanium dioxide mixed-oxide gel. Journal of Scientific and Industrial Research, Vol. 57, Issue 9. 1998. pp. 529-535.

Van Velzen, D., Langenkamp, H., and Herb, G. Antimony, its sources, applications and flow paths into urban and industrial waste: a review. Waste Management and Research. Vol. 16, Issue 1. 1998. pp. 32-40.

Varghese, O. K., Gong, D., Paulose, M., Ong, K. G., and Grimes, C. A. Hydrogen sensing using titanium dioxide nanotubes. Sensors and Actuators B: Chemical, Vol. 93, Issue 1. 2003. pp. 338-344.

Wang, H., Zhou, A., Peng, F., Yu, Hao., and Lang, J. Mechanism study on adsorption of acidified multi-walled carbon nanotubes to Pb(II). Journal of Colloid and Interface Science. Vol. 316. 2007. pp. 277-283.

Warheit, D., Laurence, B., Reed, K., Roach, D., Reynolds, G., Webb, T. Comparative Pulmonary Toxicity Assessment of Single-wall Carbon Nanotubes in Rats. Toxicological Sciences. Vol. 77. 2004. pp. 117-125.

Winship, K. Toxicity of antimony and its compounds. Adverse Drug Reactions and Acute Poisoning Reviews. Vol. 6, No. 2. 1987. pp. 67-90.

Yang, C. Statistical Mechanical Study on the Freundlich Isotherm Equation. Journal of Colloid and Interface Science. Vol. 208. 1998. pp. 378-387.

Yang, P., Lu, C., Hua, N, Du, Y. Titanium dioxide nanoparticles co-doped with Fe3+ and Eu3+ ions for photocatalysis. Materials Letters. Vol. 57, Issue 4. 2002. pp. 794-801.

Yean, S., Cong, L., Yavuz, C. Effect of magnetite particle size on adsorption and desorption of arsenite and arsenate. Journal of Materials Research. Vol. 20, Issue 12. 2005. pp. 3255-3264.

Yuan, L., Saito, K., Pan, C., Williams, F., Gordon, A. Nanotubes from methane flames. Chemical Physics Letters. Vol. 340, Issue 3. 2001. pp. 237-241.

Yuan, L., Saito, K., Hu, W., Chen, Z. Ethylene synthesis of well-aligned multi-walled carbon nanotubes. Chemical Physics Letters. Vol. 346, Issue 1. 2001. pp. 23-28.

Zhang, Y., Dai, H. Formation of metal nanowires on suspended single-walled carbon nanotubes. Applied Physics Letters. Vol. 77. 2000. p. 3015.

Zhang, Y. S., Wang, R. G., and Wang, X. X. The comparison of Cu(II) adsorption capability of baker's yeast, nano-titanium dioxide and their composite adsorbent. Chinese Science Bulletin, Vol. 53, Issue 9. 2008. pp. 1365-1372.

6.0 LIST OF TABLES

Chapter 1

Table 1.6: Physicochemical significance of Langmuir
and Freundlich parameters 27

Chapter 2

Table 2.0: Properties of materials used in experiments 30

Table 2.3: Leaching test experimental setup 32

Table 2.5: Surface oxidation experimental setup 33

Table 2.6: Comparison of MWCNTs, AC, n-TiO2
and μ-TiO2: experimental setup 35

Table 2.11a: ICP-AES operational setup 37

Table 2.11b: ICP-MS operational setup 38

Table 2.12: HPLC operational setup 39

Chapter 3

Table 3.3a: MWCNT/functionalized MWCNT
linear regression parameters 50

Table 3.3b: MWCNT/functionalized MWCNT
determined Langmuir parameters 50

Table 3.3c: MWCNT/functionalized MWCNT
determined Freundlich parameters 51

Table 3.4a: Pb, Sb, and Cd adsorption linear regression parameters 65

Table 3.4b: Pb, Sb, and Cd adsorption Langmuir parameters 66

Table 3.4c: Pb, Sb, and Cd adsorption Freundlich parameters 67

Table 3.6: Summary of Sb(III) and Sb(V) adsorption by
micro- and Nanoparticle titanium dioxide 83

7.0 LIST OF FIGURES

Chapter 1

Figure 1.2a: Armchair nanotube 9

Figure 1.2b: Zig-zag nanotube 9

Figure 1.2c: Multi-walled carbon nanotube 10

Figure 1.3a: Crystal structure of rutile 13

Figure 1.3b: Crystal structure of anatase 14

Figure 1.3c: Crystal structure of brookite 14

Figure 1.5a: Diagram of an inductively coupled plasma – mass spectrometer 20

Figure 1.5b: Ion exchange chromatography 22

Figure 1.5c: HPLC-ICP-MS coupling 23

Chapter 3

Figure 3.1a: Results of nickel leach test 41

Figure 3.1b: Nickel adsorption by pristine MWCNTs 41

Figure 3.2: Fraction of total Sb adsorption as a function of MWCNT or AC concentration 42

Figure 3.3a: Linear isotherm of the MWCNT adsorption of antimony, lead, and cadmium 44

Figure 3.3b: Linear isotherm of the surface oxidized MWCNT adsorption of antimony, lead, and cadmium 45

Figure 3.3c: Langmuir isotherm of the MWCNT adsorption of antimony, lead, and cadmium 46

Figure 3.3d: Langmuir isotherm of the surface oxidized MWCNT adsorption of antimony, lead, and cadmium 47

Figure 3.3e: Freundlich isotherm of the MWCNT
adsorption of antimony, lead, and cadmium 48

Figure 3.3f: Freundlich isotherm of the surface oxidized MWCNT
adsorption of antimony, lead, and cadmium 49

Figure 3.4a: Sb, Cd, and Pb linear adsorption
isotherms for activated carbon 53

Figure 3.4b: Sb, Cd, and Pb linear adsorption
isotherms for pristine multi-walled carbon nanotubes 54

Figure 3.4c: Sb, Cd, and Pb linear adsorption
isotherms for microparticle titanium dioxide 55

Figure 3.4d: Sb, Cd, and Pb linear adsorption
isotherms for nanoparticle titanium dioxide 56

Figure 3.4e: Sb, Cd, and Pb Langmuir adsorption
isotherms for activated carbon 57

Figure 3.4f: Sb, Cd, and Pb Langmuir adsorption
isotherms for pristine multi-walled carbon nanotubes 58

Figure 3.4g: Sb, Cd, and Pb Langmuir adsorption
isotherms for microparticle titanium dioxide 59

Figure 3.4h: Sb, Cd, and Pb Langmuir adsorption
isotherms for nanoparticle titanium dioxide 60

Figure 3.4i: Sb, Cd, and Pb Freundlich adsorption
isotherms for activated carbon 61

Figure 3.4j: Sb, Cd, and Pb Freundlich adsorption
isotherms for pristine multi-walled carbon nanotubes 62

Figure 3.4k: Sb, Cd, and Pb Freundlich adsorption
isotherms for microparticle titanium dioxide 63

Figure 3.4l: Sb, Cd, and Pb Freundlich adsorption
isotherms for nanoparticle titanium dioxide 64

Figure 3.5a: n-TiO$_2$ Sb adsorption capacity
as a result of initial Sb concentration 68

Figure 3.5b: Equilibrium Sb Concentration
Dependence on Initial Sb Concentration.. 69

Figure 3.6a: Plot of the Sb(V) adsorption capability
of nanoparticle titanium dioxide (2 mg/mL) in light. Trial 1 71

Figure 3.6b: Plot of the Sb(V) adsorption capability
of nanoparticle titanium dioxide (2 mg/mL) in light. Trial 2 72

Figure 3.6c: Plot of the Sb(V) adsorption capability
of microparticle titanium dioxide (2 mg/mL) in light. Trial 1 73

Figure 3.6d: Plot of the Sb(V) adsorption capability
of microparticle titanium dioxide (2 mg/mL) in light. Trial 2 74

Figure 3.6e: Plot of the Sb(III) adsorption capability
of nanoparticle titanium dioxide (2 mg/mL) in darkness. Trial 1 75

Figure 3.6f: Plot of the Sb(III) adsorption capability
of nanoparticle titanium dioxide (2 mg/mL) in darkness. Trial 2 76

Figure 3.6g: Plot of the Sb(III) adsorption capability
of nanoparticle titanium dioxide (2 mg/mL) in light. Trial 1 77

Figure 3.6h: Plot of the Sb(III) adsorption capability
of nanoparticle titanium dioxide (2 mg/mL) in light. Trial 2 78

Figure 3.6i: Plot of the Sb(III) adsorption capability
of microparticle titanium dioxide (2 mg/mL) in darkness. Trial 1 ... 79

Figure 3.6j: Plot of the Sb(III) adsorption capability
of microparticle titanium dioxide (2 mg/mL) in darkness. Trial 2 ... 80

Figure 3.6k: Plot of the Sb(III) adsorption capability
of microparticle titanium dioxide (2 mg/mL) in light. Trial 1 81

Figure 3.6l: Plot of the Sb(III) adsorption capability
of microparticle titanium dioxide (2 mg/mL) in light. Trial 2 82

Figure 3.7a: DRIFT spectra of pristine n-TiO2
 and n-TiO2 with adsorbed Sb, Trial 1 85

Figure 3.7b: DRIFT spectra of pristine n-TiO2
 and n-TiO2 with adsorbed Sb, Trial 2 86

Figure 3.7c: DRIFT spectra of pristine n-TiO2
 and n-TiO2 with adsorbed Sb, Trial 1 86

Figure 3.7d: DRIFT spectra of pristine n-TiO2
 and n-TiO2 with adsorbed Sb, Trial 2 87

www.ingramcontent.com/pod-product-compliance
Lightning Source LLC
Chambersburg PA
CBHW070155230526
45471CB00002B/675